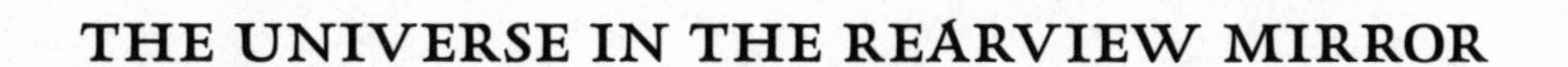
THE UNIVERSE IN THE REARVIEW MIRROR

The UNIVERSE in the REARVIEW MIRROR

HOW HIDDEN SYMMETRIES SHAPE REALITY

Dave Goldberg

DUTTON

DUTTON
Published by the Penguin Group
Penguin Group (USA) Inc., 375 Hudson Street,
New York, New York 10014, USA

USA | Canada | UK | Ireland | Australia | New Zealand | India | South Africa | China
Penguin Books Ltd, Registered Offices: 80 Strand, London WC2R 0RL, England
For more information about the Penguin Group visit penguin.com.

LIBRARY OF CONGRESS CATALOGING-IN-PUBLICATION DATA
has been applied for.

ISBN 978-0-525-95366-1

Printed in the United States of America
10 9 8 7 6 5 4 3 2 1

Set in Adobe Caslon Pro with Hadriano Std
Designed by Daniel Lagin

Art on pages xiv, xvi, xvii, 7, 22, 24, 30, 32, 50, 57, 59, 75, 81, 82, 95, 109, 116, 121, 126, 136, 139, 140, 152, 154, 160, 169, 175, 180, 183, 185, 195, 200, 207, 213, 216, 233, 242, 246, 251, 252, 273, 275, 277, and 290 courtesy of Herb Thornby

Art on page 17 courtesy of NASA, ESA, and the Hubble Heritage Team (STScI/AURA)—ESA/Hubble Collaboration; Acknowledgment: B. Whitmore (Space Telescope Science Institute)

Art on page 66 courtesy of Michael R. Blanton and the SDSS Collaboration, www.sdss.org

Art on page 93 courtesy of NASA/WMAP Science Team

Art on page 118 courtesy of NASA, Andrew Fruchter and the ERO Team [Sylvia Baggett (STScI), Richard Hook (ST-ECF), Zoltan Levay (STScI)] (STScI)

Art on page 210 courtesy of NASA, ESA, and the Hubble Heritage Team (STScI/AURA)-ESA/Hubble Collaboration

Art on page 283 courtesy of Peter McMullen

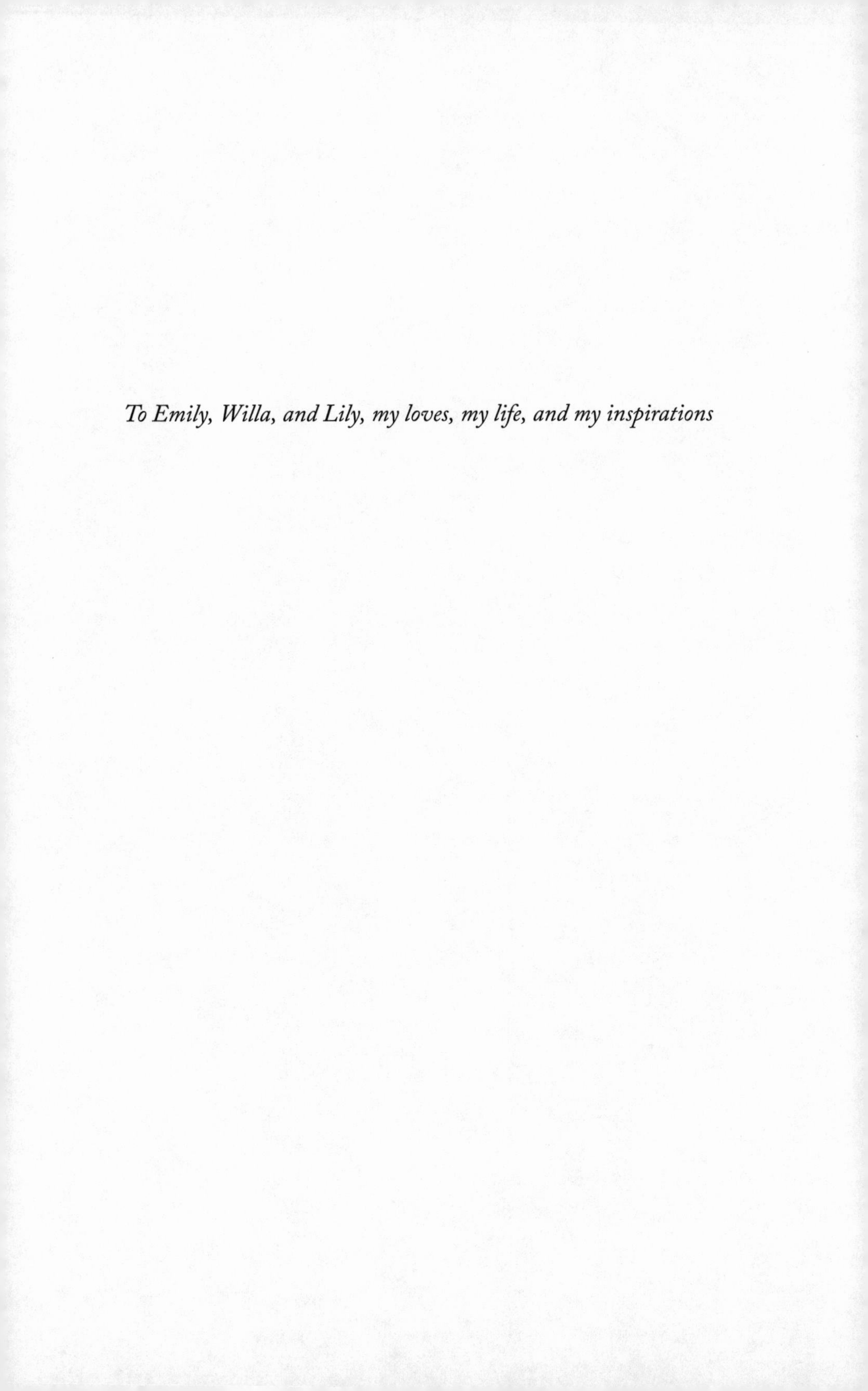

To Emily, Willa, and Lily, my loves, my life, and my inspirations

CONTENTS

We have to remember that what we observe is not nature herself, but nature exposed to our method of questioning.

—WERNER HEISENBERG

INTRODUCTION

IN WHICH I SET EVERYTHING UP, SO IT'S PROBABLY BEST NOT TO SKIP AHEAD

Why is there something rather than nothing?

Why is the future different from the past?

Why are these questions a serious person should even ask?

There is a gleeful skepticism of the orthodox in popular discussion of science. Reading some of the twittering, blogging chatter out there, you might suppose that relativity is nothing more than the ramblings of some dude at a party instead of one of the most successful physical theories ever, and one that has passed every observational and experimental test thrown at it for a century.

To the uninitiated, physics can seem littered with a ridiculous number of rules and equations. Does it have to be so complicated?

Physicists themselves sometimes bask in the aloof complexity of it all. A century ago when asked if it was true that only three people in the world understood Einstein's Theory of General Relativity, Sir Arthur Eddington thought for a few moments and casually replied, "I'm trying to think who the third person is." These days, relativity is considered part of the standard physicist toolkit, the sort of thing taught every day to students barely out of short pants. So let's put aside the highfalutin idea that you have to be a genius to understand the mysteries of the universe.

The deep insights into our world have almost never been the result of simply coming up with a new equation, whether you are Eddington or Einstein. Instead, breakthroughs almost always come in realizing that things that appear different are, in fact, the same. To understand how things work, we need to understand symmetry.

The great twentieth-century Nobel laureate Richard Feynman* likened the physical world to a game of chess. Chess is a game filled with symmetries. The board can be rotated half a turn and it will look just as it did before you started. The pieces on one side are (except for the color) a nearly perfect mirror reflection of the pieces on the other. Even the rules of the game have symmetries in them. As Feynman put it:

> The rule on the move of a bishop on a chessboard is that it moves only on the diagonal. One can deduce, no matter how many moves may be made, that a certain bishop will always be on a red square. . . . Of course, it will be, for a long time, until all of a sudden, we find that it is on a *black* square (what happened of course, is that in the meantime it was captured, another pawn crossed for queening, and it turned into a bishop on the black square). That is the way it works in physics. For a long time we will have a rule that works excellently in an over-all way, even when we cannot follow the details, and then some time we may discover a *new rule.*

* You cannot do better than read *The Feynman Lectures on Physics* except by listening to them. The quote above is taken from the recording of a lecture he delivered at Caltech. He'd intended the lectures for freshmen, but apparently by the end of term the seats had filled up with his fellow faculty.

Watch a few more games, and you might be struck by the insight that the reason a bishop always stays on the same color is that it always goes along a diagonal. The rule about conservation of color *usually* works, but the deeper law gives a deeper explanation.

Symmetries show up just about everywhere in nature, even though they may seem unremarkable or even obvious. The wings of a butterfly are perfect reflections of one another. Their function is identical, but I would feel extremely sad for a butterfly with two right wings or two left ones as he pathetically flew around in circles. In nature, symmetry and asymmetry generally need to play off one another. Symmetry, ultimately, is a tool that lets us not only figure out the rules but figure out *why* those rules work.

Space and time, for instance, aren't as different from one another as you might suppose. They are a bit like the left and right wings of a butterfly. The similarity between the two forms the basis of Special Relativity and gives rise to the most famous equation in all of physics. The laws of physics seem to be unchanging over time—a symmetry that gives rise to conservation of energy. It's a good thing too; it's thanks to the conservation of energy that the giant battery that is the sun manages to power all life here on earth.

To some people's (okay, physicists') minds, the symmetries that have emerged from our study of the physical universe are as beautiful as that of diamonds or snowflakes or the idealized aesthetic of a perfectly symmetric human face.

The mathematician Marcus du Sautoy put it nicely:

> Only the fittest and healthiest individual plants have enough energy to spare to create a shape with balance. The superiority of the symmetrical flower is reflected in a greater production of nectar, and that nectar has a higher sugar content. Symmetry tastes sweet.

Platonic Solids

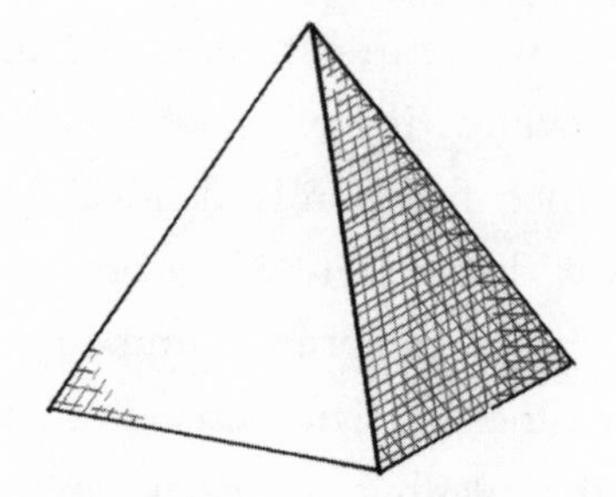

Tetrahedron
4 faces — 4 vertices

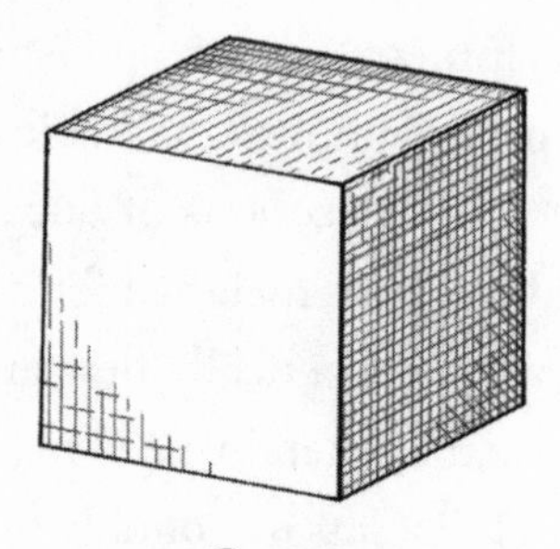

Cube
6 faces — 8 vertices

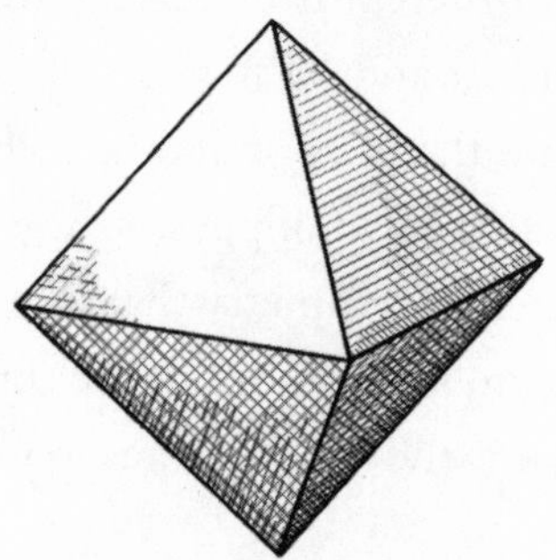

Octahedron
8 faces — 6 vertices

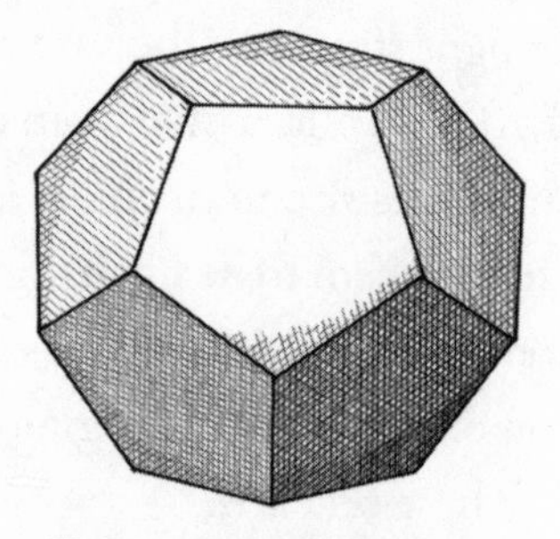

Dodecahedron
12 faces — 20 vertices

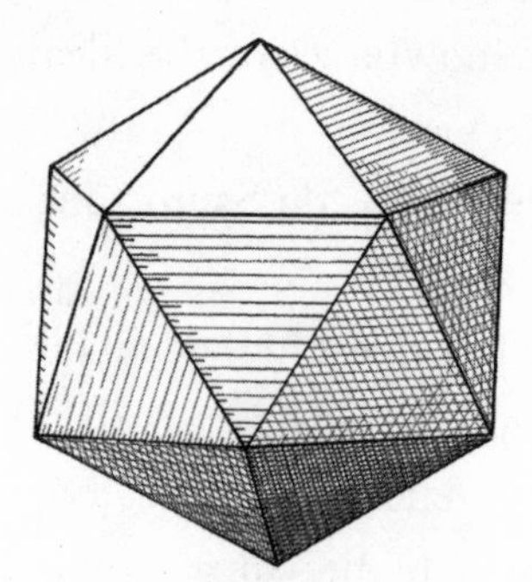

Icosahedron
20 faces — 12 vertices

Our minds enjoy the challenge of symmetries. In American-style crosswords, typically the pattern of white and black squares look identical whether you rotate the entire puzzle a half a turn or view it in a mirror. Great works of art and architecture: the pyramids, the Eiffel tower, the Taj Mahal, are all built around symmetries.

Search the deepest recesses of your brain, and you may be able to summon the five Platonic solids. The only regular three-dimensional figures with identical sides are the tetrahedron (four sides), cube (six), octahedron (eight), dodecahedron (twelve), and the icosahedron (twenty). A nerd (for example, me) will think back fondly to his early years and recognize these as the shapes of the main dice in a Dungeons & Dragons set.*

Symmetry can simply refer to the way things "match" or "reflect" themselves in our daily casual chitchat, but of course it has a much more precise definition. The mathematician Hermann Weyl gave a formulation that's going to serve us well throughout this book:

> A thing is symmetrical if there is something you can do to it so that after you have finished doing it, it looks the same as before.

Consider an equilateral triangle. There are all sorts of things that you can do to a triangle and keep it exactly the same. You can rotate a third of a turn, and it will look as it did before. Or you could look at it in a mirror, and the reflection will look the same as the original.

The circle is a symmetric object par excellence. Unlike triangles, which look the same only if you turn them a specific amount, you can

* Black belt level nerds will notice that I've somehow omitted the ten-sided die. Well, I'll have you know that the D10 is *not* a Platonic solid. It's one of a class of objects known as antidipyramids and goes by the charmingly ridiculous name of Bimbo's lozenge.

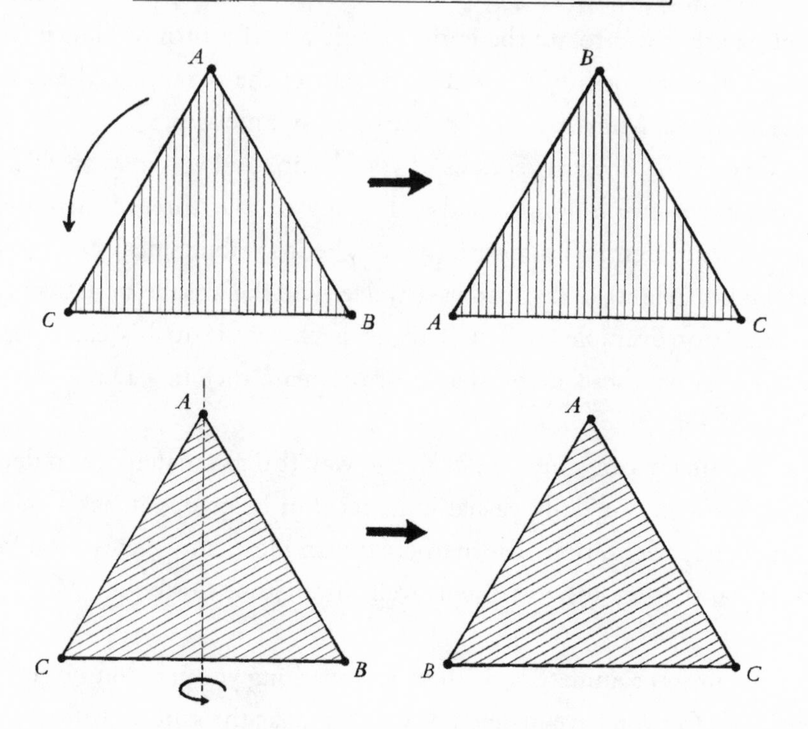

rotate a circle by any amount and it looks the same. Not to belabor the obvious, but this is how wheels work.

Long before we understood the motions of the planets, Aristotle assumed that orbits must be circular because of the "perfection" of the circle as a symmetric form. He was wrong, as it happens, as he was about most everything he said about the physical world.

It's tempting to fall into the sense of sweet, smug satisfaction that comes from mocking the ancients, but Aristotle was right in a very important sense. Although planets actually travel in ellipses around the sun, the gravitational force toward the sun is the same in all directions. Gravity is symmetric. From this assumption, and a smart guess about how gravity weakens with distance, Sir Isaac Newton correctly deduced the motions of planets. This is one of the many reasons you know his name. Something that doesn't look nearly as perfect as a circle—the elliptical orbits of the planets—is a consequence of a much deeper symmetry.

Symmetries reveal important truths throughout nature. An understanding of how genetics really worked had to wait until Rosalind Franklin's x-ray imaging of DNA allowed James Watson and Francis Crick to unravel the double helix structure. This structure of two complementary spiral strands allowed us to understand the method of replication and inheritance.

Double Helix

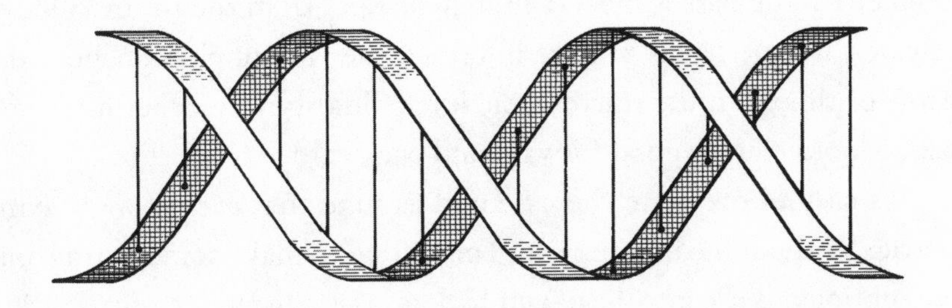

If you run in particularly geeky circles, you may have heard a scientist refer to a theory as *natural* or *elegant*. What this normally means is that an idea is based on assumptions so simple that they absolutely *must* be correct. Or to put it another way, that if you start with a very simple rule, you could derive all sorts of complicated systems like the gravity around black holes or the fundamental laws of nature.

This is a book about symmetry: how it shows up in nature, how it guides our intuition, and how it shows up in unexpected ways. The Nobel laureate Phil Anderson put it most succinctly:

It is only slightly overstating the case to say that physics is the study of symmetry.

Some symmetries will be so obvious as to seem to be completely trivial, but will produce some incredibly nonintuitive results. When you ride on a roller coaster, your body can't distinguish between being pushed into your seat by gravity or by the acceleration of your car; the two feel the same. When Einstein supposed that "feels the same" really means "*is* the same," he derived how gravity really works, eventually leading to the proposal of black holes.

Or the fact that you can swap two particles of identical type will lead, inexorably, toward an understanding of the fate of our sun and the mysterious Pauli Exclusion Principle, and ultimately to the functioning of neutron stars and all of chemistry.

The flow of time, on the other hand, seems to be just as obviously *not* symmetric. The past is most definitely distinct from the future. Oddly, however, no one seems to have informed the laws of physics about the arrow of time. On the microscopic level, almost every experiment you can do looks equally good forward and backward.

It's easy to overstate the case and assume that *everything* is symmetric. Without having met you, I'm willing to make some outrageous assumptions. Back in college you had at least one stoner conversation along the lines of "What if our whole *universe* is just an atom in a way bigger universe, man?"

Have you grown up any since then? Admit it, you saw the perfectly decent *Men in Black*, or think back fondly to your childhood reading *Horton Hears a Who!*, and even now, you can't help but wonder if there is a miniature universe far beyond our perceptions.

The answer, Smoky, is no, but the *why* is a somewhat deeper question.

If you can make something bigger or smaller without changing it, you are demonstrating a particular type of symmetry. For those of you who've read *Gulliver's Travels*, you may recall that when we meet the Lilliputians,* Jonathan Swift goes into excruciating detail explaining the consequence of the difference in height between Gulliver and the Lilliputians and, later, between Gulliver and the giant Brobdingnags. Swift really belabors the point, describing the ratios of everything from the size of a man's step to the number of local animals required to feed Gulliver.

But even in Swift's time, it was pretty well established that the story wouldn't make physical sense (to say nothing of the talking horses). A hundred years earlier, Galileo wrote his *Two New Sciences*, in which he probes the scientific plausibility of giants.† After much deliberation, he concludes against the proposition, basically ruining everyone's fun forever. The problem is that a bone that doubles in length becomes eight times heavier but has only four times the surface area. Eventually it would collapse under its own weight. As he puts it:

> An oak two hundred cubits high would not be able to sustain its own branches if they were distributed as in a tree of ordinary size; and that nature cannot produce a horse as large as twenty ordinary horses or a giant ten times taller than an ordinary man unless by miracle or by greatly altering the proportions of his limbs and especially his bones, which would have to be considerably enlarged over the ordinary.

He obligingly sketches a giant's bones for the benefit of the reader. And concludes with the adorably disturbing imagery:

---

* The Lilliputians are one-twelfth the size of Gulliver in every dimension. Multiplying by ten is way easier, so for my bridges and whatnot, I decided to round and make things a bit simpler. You're welcome.

† Yeah. That seems like a good use of his time and talents.

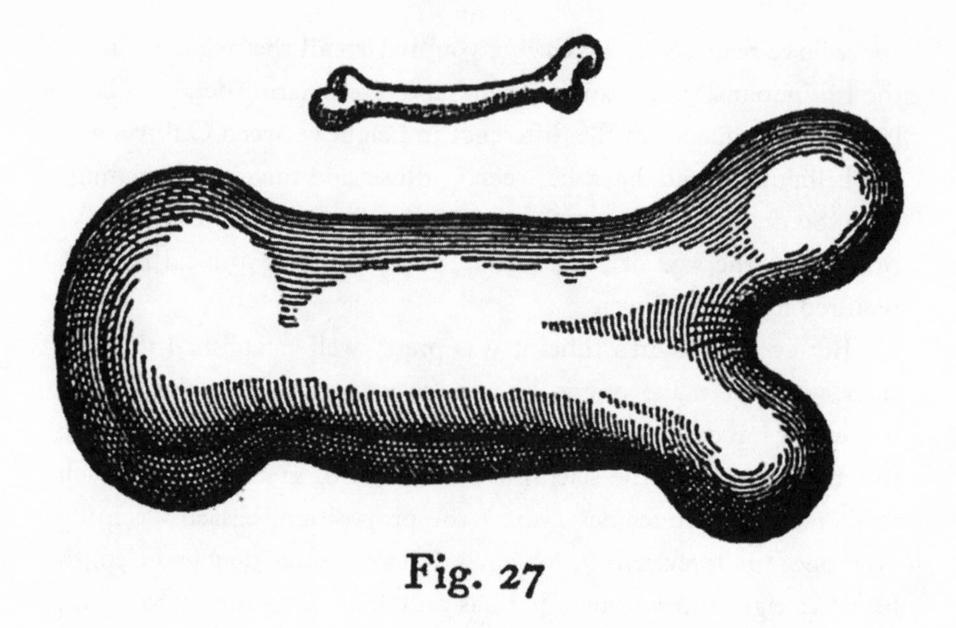

Fig. 27

> Thus a small dog could probably carry on his back two or three dogs of his own size; but I believe that a horse could not carry even one of his own size.

This is why Spider-Man is such an ill-conceived premise.* Spidey wouldn't have the proportional strength of a spider. He'd be of such bulky construction that he wouldn't even need squashing. Gravity would do the trick for you. As the biologist J. B. S. Haldane puts it in his essay "On Being the Right Size":

> An insect, therefore, is not afraid of gravity; it can fall without danger, and can cling to the ceiling with remarkably little trouble. . . . But there is a force which is as formidable to an insect

* It is a well-established fact that if you talk to a science nerd for long enough, they will ruin *everything* by looking into it too deeply. This is why we spend so, so many nights alone.

> as gravitation to a mammal. This is surface tension. . . . An insect going for a drink is in as great danger as a man leaning out over a precipice in search of food. If it once falls into the grip of the surface tension of the water—that is to say, gets wet—it is likely to remain so until it drowns.

The problem goes much deeper than just the tensile strength of the bones of giants and the proportional strength of insects. Although things on the human scale seem to scale up or down fine—a 20-foot killbot seems like it'd work twice as well and with the exact same design as a 10-foot model—once you get down to atomic scales, all bets are off. The atomic world is also the quantum mechanical world, and that means that the concreteness of our macroscopic experience is suddenly replaced with uncertainty.

Put another way, the act of scaling is *not* a symmetry of nature. A map of the cosmic web of galaxies kind of looks like a picture of neurons, but this isn't some grand symmetry of the universe. It's just a coincidence.

I could go on describing symmetry after symmetry, but I trust I've made the point. Some changes matter and some don't. My approach in this book is to focus each chapter on a specific question that will turn out to be answered, however indirectly, by fundamental symmetries in the universe.

On the other hand, that other hand isn't the same as its partner. One of the most important puzzles humans can ever ponder is that, in some sense, the universe *isn't* symmetric. Your heart is on the left side of your chest; the future is different from the past; you are made of matter and not antimatter.

So this is also, or perhaps more fundamentally, a book about broken and imperfect symmetries. There's a proverb: A Persian rug is perfectly imperfect, and precisely imprecise. Traditional rugs have a small imperfection, a break in the symmetry that gives the whole thing more character. So too will it be with the laws of nature, and a good thing because a perfectly symmetric universe would be staggeringly boring. Our universe is anything but.

The universe in the rearview mirror is closer than it appears—and that makes all of the difference in the world. But let's not look back; we're on a tour of the universe. Symmetry will guide our way, but symmetry breaking will make our tour something to write home about.

Chapter 1

ANTIMATTER

IN WHICH WE LEARN WHY THERE IS SOMETHING RATHER THAN NOTHING

It is generally a bad idea to watch science fiction in the hopes of bolstering your understanding of science. Doing so would give you a very distorted impression of, among other things, how explosions sound in deep space (they don't), how easy it is to blast past the speed of light (you can't), and the prevalence of English-speaking, vaguely humanoid, but still sexy, aliens (they're all married). But if we've learned one good lesson from *Star Wars*es and *Trek*s, it's that no one should ever mess with antimatter.

The awesome power of antimatter is irresistible, and a writer trying to interject some gen-u-ine science into the mix will often reach for a handful of antimatter to make themselves seem respectable. The *Enterprise* has, at its heart, a matter/antimatter engine. Isaac Asimov used a positronic brain in his robots—turning an antimatter particle, the positron, into a sci-fi MacGuffin.

Even Dan Brown's *Angels & Demons*—hardly hard sci-fi in the normal sense of the term—uses antimatter as a sort of doomsday device. His villains steal half a gram of antimatter, enough to produce an explosion comparable to early nuclear bombs. Except for a factor-of-two calculation error on Brown's part,* a complete misunderstanding of what particle accelerators actually *do*, and overestimating by roughly a factor of a trillion how much antimatter could be stored and transported, he gets the science part exactly right.

But despite our exposure to it, antimatter is still terribly misunderstood. It's not the unstoppable killing material you've grown to distrust all of these years. Left by itself, it's quite benign. Antimatter is just like the regular stuff you know and love—the same mass, for instance—but reversed: reversed charge and reversed name. It's only when you start mixing antimatter with normal matter that things get explode-y.

Antimatter is not only no more exotic than ordinary matter but, in almost every way that matters, it looks and acts the same. Were every particle in the universe to suddenly be replaced by its antimatter version, you wouldn't even be able to tell the difference. To put it bluntly, there is a symmetry between how the laws of physics treat matter and antimatter, and yet they must be at least a *little bit* different; you and everyone you know are made of matter and not antimatter.

We like to think accidents don't happen, that there is some ultimate cause to explain why, for instance, you're not standing around in a room full of anti-people. To understand why that is, we're going to delve into the past.

NEVER MIND ANTI-PEOPLE, WHERE DID *I* COME FROM?

Origin stories are tough. Not everything can be explained as neatly as being bitten by a radioactive spider, having your home planet explode, or even the reanimation of dead tissue (you know, for science). Our own origin story is complicated, but you'll be pleased to know that, much

* When antimatter goes *boom!* it takes out an equal amount of matter. Brown apparently forgot that part.

like *The Incredible Hulk*, we're also ultimately the result of exposure to gamma radiation. It's a long story.

Physics can't yet answer the question of where the universe itself came from, but we can say a lot about what happened afterward. At the risk of triggering an existential crisis, we can at least try to answer one of the biggies in the philosophical pantheon: Why is there something rather than nothing?

This isn't as dumb a question as it would seem. Based on everything that we've ever seen in a lab, you should not exist. It's nothing personal. I shouldn't exist, either, nor should the sun, the Milky Way Galaxy, or (for many, many reasons) the *Twilight* movies.

To understand why you *shouldn't* exist, we're going to have to look into mirror universes, antimatter universes, and our own universe on the smallest scale. It's only at the most microscopic scales that the differences between matter and antimatter show up, and even then, they are far from obvious.

The universe on the smallest scales is *different*.* Everything we can see is made of molecules, the smallest of which are about a billionth of a meter across. Put into human scales, a human hair is the equivalent of about a hundred thousand molecules across. Molecules are *small*, but small as they are, molecules aren't the most basic things in the world. It's a good thing too, at least if we're interested in making any real order out of things. According to the Royal Society of Chemistry, there are more than 20 million different kinds of known molecules, and more are being discovered so quickly that it's completely pointless to try to pinpoint an exact number. Without the realization that molecules are made of something smaller, we'd be stuck just listing them.†

---

* I hope you didn't skip the intro. There's a lot of good stuff in there.

† Ernest Rutherford, who did as much as anyone to explain the structure of matter, ungraciously observed, "All science is either physics or stamp collecting." It must have really sucked for him when, in 1908, he won the Nobel prize for *chemistry*.

Fortunately for the sake of universal order, new structures appear as we probe smaller and smaller scales. On a scale of less than a ten billionth of a meter we start to see individual atoms. There are only 118 known atomic elements, the majority of which don't occur naturally or are found only in trace amounts.

Nothing we see on a macroscopic scale really prepares us for what we see when we get to the size of atoms, because that is when quantum mechanics kicks in. I'm going to hold off for a bit on dealing with the quantum nature of reality, but it can get disturbingly uncertain. We can ignore it for now, but it's a morass we're going to have to wade neck deep into later.

Even without knowing exactly what atoms *are*, it's possible to make some sense of them. This is precisely what the Russian chemist Dmitri Mendeleev* found in the nineteenth century. You're probably familiar with his work if you've ever been in a science classroom. Mendeleev invented the periodic table.

It's not just a big list. Mendeleev showed that the elements in various columns have very similar chemical properties to one another. Copper, silver, and gold, for instance, are all in the same column, and all of them are metals with pretty good conductivity. By filling in the blanks, Mendeleev was able to predict the properties of elements before they were even discovered in the lab.

The idea of an atom as the indivisible basis of all matter has been around for two and a half millennia, albeit in a fairly primitive form. Leucippus, Democritus, and the ancient Greek "atomists" came up with the idea in the fifth century BCE, and we might easily suppose that

---

* Mendeleev is the first of many people we'll talk about in this book who were essentially robbed of a Nobel Prize. In his case, political maneuvering denied him the prize in chemistry 1907, despite the fact that the periodic table is the foundation for all of modern chemistry and atomic physics.

we've spent the last few millennia playing catch-up. Personally, I think this is giving the ancients far too much credit.

The original atomists basically said that you couldn't keep slicing up matter indefinitely. They had no idea how small atoms were, how they were structured, or whether (despite the word *atom* literally meaning "indivisible") they could actually be broken up further.

It's only been in the last couple of centuries that we've really had any understanding of what atoms were really like, culminating with Einstein's brilliant analysis of Brownian Motion in 1905. Eighty years earlier, the botanist Robert Brown studied pollen suspended in a liquid under a microscope. Brown noticed that no matter how long he waited for things to settle down, the grains continuously jittered around in a random motion.

Einstein reasoned, rightly, that individual molecules were continuously jostling the grains in random directions, and from that, he was able to figure out that atoms were very much real, and approximately how large they were.

The conclusive discovery of atoms *should* have been more than sufficient to make Einstein one of the greatest scientists of the twentieth century, but it was arguably only the third most important thing that he did in the space of a single year.

In what can only be considered the greatest concentration of brilliant scientific output ever, Einstein spent 1905—his Miracle Year as it's come to be known—publishing a series of papers in which he not only demonstrated the existence of the atom but also showed that light is composed of particles (for which he won the Nobel Prize in 1921) and invented a little something called relativity—the main reason you probably know his name.

Particles may seem like a Platonic abstraction at first. They are fundamental and indivisible. They have no shape, size, color, or any other macroscopic properties. And any particle of a type will be completely identical and indistinguishable from one of the same type. Quite literally, if you've seen one electron, you've seen them all.

Knowing *that* atoms are, doesn't necessarily tell you *what* atoms are or even if they are fundamental particles. (A: They aren't.) To solve that mystery, we turn to Ernest Rutherford, who, in 1911, was busy shooting alpha particles (a sci-fi-esque name for what we now know were helium nuclei) into a sheet of gold foil.

There's no need to go on and on about all the wrong turns that science took before we reached the (good but obviously not perfect) model we have today, but before Rutherford, people really had no idea how an atom was structured. The prevailing idea was that the whole thing was filled with a "pudding" of positive charge, with little "plums" (electrons) scattered throughout.* Electrons are probably already familiar to you. They were the first elementary particle to be discovered, way back in 1897 when J. J. Thomson referred to them (disgustingly) as "corpuscles."

Electrons are also incredibly easy to make: Just take a piece of metal, heat it up, and electrons come flying off! Or, if you're not allowed to turn on the stove by yourself, just wear wool socks and touch a metal surface. That shock? That's science! (also, electrons).

If the Plum Pudding Model were correct, Rutherford's alpha particles would be slightly deflected as they went through the pudding. Instead, most of the alpha particles went through completely unscathed, while a few came flying backward, as if they'd hit something solid. In Rutherford's words:

> It was quite the most incredible event that has ever happened to me in my life. It was almost as incredible as if you fired a 15-inch shell at a piece of tissue paper and it came back and hit you.

---

* Apparently, plum pudding was a treat around the turn of the twentieth century. Personally, I think it sounds nauseating.

Rutherford Scattering

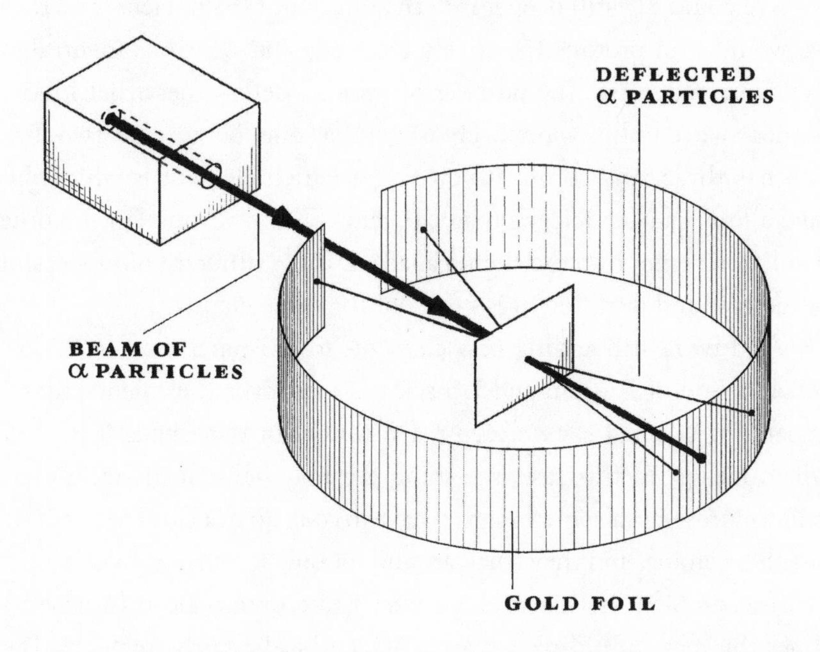

Most of the alpha particles went clear through. It was only an especially rare one that happened upon a gold nucleus. In other words, the vast, vast majority of the mass of atoms was confined to a tiny fraction of the total volume. There were plums, but no pudding.

You may think of your hand as being solid and full of stuff but for the most part, it's really empty space. We'd have to zoom in by a factor of 100,000 smaller than the atom itself (to scales of 10^{-15} meter) before we could see the atomic nucleus, and here we realize how empty our existence is.

The nucleus makes up about 99.95 percent of the mass of an atom, but takes up only about 1 quadrillionth of the total volume. It's something like the ratio of a modest-size office building compared to the entire earth. Each alpha particle in the Rutherford experiment is roughly equivalent to a single meteor striking the earth and—just by random chance—hitting the White House.* Most are going to miss.

We could dig still deeper into the interior of the nucleus, and there we would find protons (positively charged) and neutrons (neutral, as the name suggests). The number of protons determines what kind of element we're talking about. Hydrogen has one, helium has two, lithium has three, and so on. If you forget which element is which, just take a look at Mendeleev's magical chart. The neutrons, on the other hand, don't enter into the chemical accounting; different numbers simply correspond to different isotopes of the same element.

And we're still adding new elements to our repertoire. In 2006, a collaboration of Russian and US scientists discovered element 118, ununoctium. When I say *discovered*, I really mean they made it in a lab, which means, in this case, smashing the holy hell out of calcium and californium—itself something you have to cook up in a lab. The result was just three atoms, and they stuck around for only the merest of instants.

The problem is that massive nuclei like ununoctium (nearly 300 times the mass of ordinary hydrogen) tend to be fairly unstable. They want to decay into lighter particles as quickly as possible Ununoctium lasts only about a millisecond, which means that you're not going to find it just lying around.

Radioactive decay is just a fact of life in the subatomic world, and it probably brings to mind nasty materials like plutonium and uranium. And to understand why these elements are so unpleasant, we're going to back away from the microscope and take a short digression to the most famous equations in physics.

---

* Alien invasion movies might give you a rather skewed impression about how likely this is.

HOW DO YOU MAKE SOMETHING OUT OF NOTHING?

No matter how much you detested your high school physics class, no matter how much of a math-phobe you might be, I'll bet there's one equation you know already, at least by reputation:

$$E = mc^2$$

Remember Einstein's Miracle Year of 1905? This equation is the jewel of his Theory of Special Relativity, and the equation at the heart of nuclear energy and our sun. And your stuff.

Everyone has at least a passing familiarity with mass and energy (m and E, respectively). What connects them is c, the speed of light and the ultimate speed limit in the universe.

To be honest, the speed of light is not terribly well named, because *any* massless particle will travel at c. There's the photon, of course, the particle of light, but there are at least a few others. Gluons, for example, are the particles responsible for holding your protons together.

Photons and gluons have a *lot* in common. There are four fundamental forces that guide the physical universe, and every one of them has at least one mediator particle that serves as a sort of subatomic middleman. This is how forces actually work. Mediator particles are the subatomic equivalents of passing a note in gym class, and for electromagnetism, photons are the conduits that tell opposite charges to repel and like charges to attract. Gluons play the same role in the strong nuclear force—the mightiest of all of the fundamental forces.

At the other extreme, gravity—surprisingly, based on everyday experience—is the *weakest* of the fundamental forces and may or may not have an associated particle. We've proactively named the *graviton* because it would be just so neat and elegant if gravity behaved like the other three fundamental forces. But we haven't actually observed a graviton yet.

All of these, photons, gluons, and gravitons (if they exist), are massless, and therefore travel at the speed of light. As you're probably made of massive particles, you're doomed forever to travel slower than light. Sorry.

Fortunately, for most purposes, this is not an onerous limit. The speed of light is very fast, around 300,000 kilometers per second, or 680 million miles per hour (mph).* It takes the earth, traveling at over 100,000 mph, 1 year to travel around the sun. It would take light only about 52 minutes to make the same trip.

Einstein's equation is a sort of currency exchange between mass and energy. You put a certain amount of mass in, destroy it entirely, and get a bunch of energy out. Lest you think I'm giving away nuclear secrets for free, this is much harder to do than it seems.

Start with a kilogram of hydrogen and raise it to something like 10 million Kelvin† and pack it tight. Congratulations! You've built yourself a nuclear fusion reactor just like the sun, capable of turning hydrogen into helium and a few other, much lighter particles.

Postfusion, you end up with about 993 grams of output, essentially the ash of your nuclear reactions. On the other hand the "missing" 7 grams is where the magic happens. It gets converted into pure energy, and though that may seem like a relatively paltry return, the hugeness of *c* means that an enormous amount of energy is released in the process. Even with the 0.7 percent conversion factor, the sun will burn‡ for

---

* Literally, *c* is an abbreviation for *celeritas*, meaning "swiftness." This is a bit of an understatement.

† The Kelvin scale counts up from absolute zero, which is the same as –273°C or –460°F. Room temperature is about 310K, and the surface of the sun is about 5,800K.

‡ Sticklers will complain about my use of the word *burn*. Technically, burning is a chemical process, and nuclear fusion is different. So sue me.

a total of about 10 billion years. A similar, coal-burning sun would last only a pathetic 10,000 years or so.

Or take the case of radioactive decay. A lump of radium will quickly start decaying into things like radon and helium, which, combined, are quite a bit lighter than the original radium atoms. That extra gets converted into high-energy x-rays.

As one of its principal discoverers, Marie Curie had to learn about the dangers of radioactivity the hard way. One of the severe hazards of the job was that she had to work in close proximity to materials like radium. She died of cancer due to her exposure, and to this day, her papers are kept in lead-lined boxes, still too radioactive for people to handle safely. Radioactive emission from her fingers is recorded on photographic film inserted in the pages.

$E = mc^2$ has another side to it. Just as you can use mass to make energy, energy can make mass as well. However, c^2 is a whopping huge number, which means that you don't get much mass out of energy under normal circumstances. But if you have enough energy, you can make some remarkable things. Suppose some particularly enterprising supercivilization wanted to make a lot of mass in a hurry. If you covered the entire earth with perfectly efficient solar panels you could, in principle, make about 2 kilograms of material every second. This would also correspond to roughly 50,000 times the total current energy consumption of all of humanity.*

We don't need to wait for a supercivilization; energy is converted to mass at the subatomic level all the time. Protons and neutrons are fairly basic building blocks of matter, but they're made of yet more fundamental particles, known as quarks, three for each. It's astonishing that if you were to add up the masses of the quarks, you'd find that they make up only about 2 percent or so of the mass of a proton. The other

* Just my little pitch for renewable energy. There's a *lot* of it out there.

98 percent comes from the insane energies involved as the quarks fly around and interact inside the protons.

To summarize: You are made of fundamental particles, which are almost entirely empty space, and the tiny bits that aren't empty space aren't all that massive. Ephemeral energy just makes them appear that way. Particles can be created from whole cloth and energy, and destroyed just as quickly. You are not just much more than the sum of your parts; strictly speaking, your parts add up to a small pile of matchsticks in a tornado of pulsing, screaming energetic interactions. *Yippee-ki-yay!*

WHERE ARE ALL THE ANTI-PEOPLE?

Energy can be used to make matter from whole cloth, but as a side effect, antimatter gets made as well. I've referred to antimatter by its effect, but haven't really said what it is. Prepare to be underwhelmed!

Every type of particle has an antimatter version that behaves almost exactly the same—the same mass, for instance—but has the opposite charge. A positron behaves just like an electron, but has a positive charge rather than a negative one. An antiproton has a negative charge, contrary to a proton's positive one, and so on.

One of the craziest things about antimatter is that if you were smart enough—and apparently only the French physicist P. A. M. Dirac was—you could have actually *predicted* antimatter before it was ever discovered. In 1928, Dirac derived the equations of relativistic quantum mechanics. Yes, that's exactly as difficult as it sounds. Plugging through the equations, he noticed that there were missing solutions. He found, for example, that electrons should pop out of the theory naturally, but other particles with the same mass and opposite charge should also be allowed.

Dirac's equation predicted that for every particle like an electron, there was going to be an antiparticle. At first, he didn't have it quite right. He thought of the positron as:

> An electron with *negative energy* [that] moves in an external field as though it carries a positive charge.

Dirac didn't know quite what his equations were saying. If his original gut reaction were correct, then you'd essentially be able to generate nearly infinite energy just by producing positrons. It would be the equivalent of running a business by running up literally infinite interest-free debt.

But ultimately, Dirac hit on the truth: Positrons are just the flip side of electrons. In other words, there seemed to be a deep symmetry between matter and the as-yet undiscovered antimatter.

Confirming this deep symmetry of nature was more than just a matter of slogging through the math. At the time, there was no experimental evidence for anything like a positron or any other antiparticle, so it was tremendously gratifying when Carl Anderson eventually discovered positrons at his lab at Caltech in 1932. Sometimes all of this high-level math turns out to have something to do with reality.

And the reality of a particle's antimatter evil twin is that while opposites may attract, it's not always such a good idea for particles and antiparticles to act on those urges. When an electron and a positron come into contact with one another, the resulting conflagration completely annihilates them both, and in the process, the magic of $E = mc^2$ turns their mass into a huge amount of energy.

There's nothing special about which particle we choose to call the "antiparticle" and which one is "normal." In a parallel universe made entirely of what we call antimatter, those anti-people would no doubt call their atoms ordinary and we'd be the anti-ones. And this is really one of those cases where both the anti-people and we are right. It's all just a matter of semantics.

That isn't to say that there's *no* antimatter in our universe. Antimatter is made all the time in the sun, which produces positrons as a side effect of fusing hydrogen into helium. Closer to home, we're able to make all sorts of exotic antiparticles in huge accelerators like the Large Hadron Collider in France and Switzerland.

It's even possible, in a lab environment, to make antimatter versions of atoms. In 2002, the European Organization for Nuclear Research (CERN)* was able to create and trap literally thousands of antihydrogen atoms with properties identical to their ordinary matter counterparts. In 2011, a new mass record was set when the Relativistic Heavy Ion Collider on Long Island created the first antihelium nucleus. In any event, antiparticles aren't long for this world. They quickly decay or collide with ordinary particles, annihilating in the process.

Sure the antimatter versions look just like ordinary particles, but are they *really*? This is our first official symmetry, and I'm going to give it a boldfaced definition just so you know it's important:

> **C Symmetry:** The physical laws for antiparticles behave *exactly the same* as for their respective particles.

Even though we give the symmetry a name (it is called C Symmetry for "charge conjugation"), that doesn't necessarily mean that matter and antimatter *really* behave the same in our universe. It's more of an educated guess.

But because we can't do the sort of magic necessary to actually swap every particle for its antiparticle, we're going to have to do a bit of speculating, which brings us to the central mystery with matter and antimatter.

In the lab—indeed, in every subatomic reaction we've ever seen—one can't create particles without creating the exact same number of antiparticles at the same time. We're able to detect elusive particles like the Higgs boson not by seeing the particle itself but rather because it decays into a particle–antiparticle pair.

---

* Known as CERN, an acronym based on the French name for the organization. These are the good folks who are also running the Large Hadron Collider (LHC).

On the flip side, if you stir an electron and a positron into a pot, you destroy both in the process, liberating all of the energy that Einstein promised you. This sort of thing happens in the vacuum of space all the time. Particles and antiparticles get created and destroyed in perfect concert.

At least that's what happens now. At some point in the distant past, matter gained the upper hand. This isn't true just for our little corner of the universe, but seems to be the case *everywhere.* One of the major, as yet still unrealized, goals of modern cosmology is to understand how the apparent C Symmetry of the universe was broken, and to do *that*, we need to look into the past.

In 2001, NASA launched the Wilkinson Microwave Anisotropy Probe (WMAP). As the acronym implies, WMAP was sent up to make a detailed map of the Cosmic Microwave Radiation, the remnant light from the early universe.

I've already said that light is ultimately made of particles known as photons, but what I glossed over is what distinguishes one photon from another. The differences ultimately come down to energy. Blue light, for instance, has a higher energy per photon than does red light. At lower energies than even red light, beyond the sensitivity of our eyes, we find the infrared, and at lower energies still we find microwaves. At the other end of the spectrum, at energies slightly too high to be detected by our eyes, there are ultraviolet photons. Beyond that, we get x-rays, and ultimately, at the very highest energies, gamma rays (γ-rays).

If you've ever worn a pair of infrared goggles, you may have noticed that living, warm-blooded creatures glow ever so slightly brighter than their cooler environment. This is why the Predator is such an effective hunter. All hot bodies give off radiation, some much more than others . . . if you know what I mean. Hot coals glow red, but the universe is much cooler than coal, around 2.7K, and glows in the microwave range. It is very, very cold in deep space.

But space wasn't always freezing. The universe is expanding, which means that the energy is becoming more and more diffuse. Early in the history of the universe, everything was packed much more tightly, and temperatures were *much* higher. For instance, 14 million years after the beginning, the universe was a balmy 310K (room temperature) and the universe glowed in the infrared. Going back even earlier, to 1 second after the Big Bang, the universe was about 10 billion degrees; a microsecond after the Big Bang, temperatures were more than 10 *trillion* degrees.

In those early days, there was plenty of energy to go around, and particle–antiparticle pairs of all sorts were constantly created. Two incredibly high-energy γ-ray photons flew into one another and *blam!* their energy was turned into an electron and a positron, or some other particle and antiparticle. See? I told you that you owed your existence to gamma rays.

The universe cooled, eventually to the point at which new pairs couldn't be created anymore. Because the universe was unable to make any new material, all of the particles and antiparticles should have eventually found and annihilated one another.

This is the key to the big mystery: If matter and antimatter are always created and destroyed in equal quantity, then there shouldn't be any of either around today, and yet here we are, all made of matter, apparently in direct contradiction to everything we've seen in the lab. It's like the universe somehow had an ace up its sleeve.

So where did you come from? And where are those anti-people?

HOW MATTER AND ANTIMATTER ARE THE SAME . . . OR NOT

We may get a few antiparticles here and there from cosmic rays or in labs, but those don't last long. Over the long haul, we're made of 100 percent matter. What happened to the symmetry that we were so excited about?

One possibility—and a science fiction writer is working on this somewhere—is that the universe really *is* symmetric in regard to matter and antimatter. Maybe half the galaxies in the universe are made of matter and the other half antimatter, and we just happen to live in a matter one.

Nice try, Captain Science, but no.

There are several problems, not least of which is the astronomical improbability of the matter and antimatter being so thoroughly segregated in the first place. This is like the left side of your coffee being steaming hot while the right side is freezing cold. What are the odds of that? To get, by merest chance, a galaxy made entirely of matter when there's an equal amount of antimatter floating around the universe is equivalent to flipping a fair coin something like 10^{69} times in a row and having it come up heads every time.

Besides, galaxies collide with each other all the time, and we have never seen an extragalactic collision with the sort of sheer, unbridled energy that would erupt if a matter galaxy rammed into an antimatter one. In short, our entire visible universe seems to be made of matter.

I now have to make a confession. We (that is, physicists) don't know why there's this imbalance, why the universe is made of matter. When it comes to matter and antimatter, the laws of physics are kind of like parental love: They *say* they love all of their kids equally, but their actions paint a very different picture. There's something special about us that prevented us from getting swept away along with the antimatter.

The grim fact is that the universe is in a constant state of decay, and generally speaking, if particles *can* decay into something lighter, they *will.* Free neutrons, for example, decay into their (slightly lighter) proton counterparts after about 10 minutes or so. But protons don't have anything to decay into.

If the symmetry between matter and antimatter is absolute, then a proton literally can't decay *ever.* And, experimentally, that may well be the truth. The current limits are that the proton lifetime is *at least* 10^{34} years. Yes, I realize that this is much, much longer than the actual age of the universe, but we're able to monitor many, many protons at once—whole swimming pools of them, in fact—and the longer we wait, the longer the minimum proton lifetime.

Here's where it gets weird. If protons never decay, then the universe really shouldn't be able to change the net balance between antimatter and matter, but if *that's* the case, then we shouldn't have ended up with an excess of matter in the first place.

That "in the first place" must have been a *very* short time after the Big Bang, and if the matter and antimatter differed a little bit back in the beginning, then presumably it does so now. If we wait long enough, eventually a proton should decay into *something*, and the *How long?* and *Into what?* will tell us a lot about the difference between matter and antimatter. To figure out the real differences between matter and antimatter, we're going to have to delve into a parallel antiuniverse, and we're not going to want to do that on our own.

If you've ever read the works of Lewis Carroll, you may have noted that the guy was *obsessed* with symmetry. In his everyday life as Charles

Dodgson, he was a mathematician, after all. We're going to need an avatar to explore the difference between our world and others that are very similar, but for simple changes like turning all matter into antimatter, I can think of no better choice than Alice (of *Alice's Adventures in Wonderland* and *Through the Looking-Glass* fame).

Antimatter Wonderland, strange as it is, will be a world identical to our own, but made entirely of antimatter. As Alice jumps down the rabbit hole into a world of antimatter, what would she find? Would she even notice?

As I've told it so far, this would be a very short (but fairly exciting) story. The moment she touched the ground—indeed, the moment she came into contact with the air—she'd get obliterated as all of her protons and neutrons annihilated with the antiprotons and antineutrons in Wonderland.

But suppose that stepping through the rabbit hole turned Alice's atoms into antiatoms as well. Is there any experiment, any experiment at all, to tell her that she's now in an antimatter Wonderland rather than a matter one? Alice isn't going to blow up or anything dramatic like that; if both Alice and the world around her are made of antimatter, she's perfectly safe.

She could build a lab, and just about every experiment will look exactly the same on either side of the portal. In the normal earth, the north poles of two bar magnets will repel one another, and the north and south poles will attract. In antimatter Wonderland, north goes to south and vice versa, but since both magnets get switched, the same ends still repel one another.

I'll save her the effort. *Nearly* any experiment that Alice can conduct will look the same in Wonderland as it did in the original. But if she is persistent enough, there's going to be one very subtle difference, and it involves an often overlooked particle known as a neutrino.

They may be overlooked, but neutrinos (which translates adorably to "little neutral ones") are among the most abundant particles in the universe. Only photons are more numerous. We just tend to ignore them because (1) they're so incredibly light that it wasn't until 1998 that the Super-Kamiokande experiment in Japan found that neutrinos had any mass at all and (2) they're electrically neutral, which means that light doesn't interact with them.

Neutrinos are incredibly hard to detect. It wasn't until 1956, and the advent of the nuclear age, that we were able to see them at all. Nuclear reactors make a lot neutrinos and antineutrinos during the course of their normal operation. Frederick Reines and Clyde Cowan, both at Los Alamos National Laboratory, set up an experiment in which antineutrinos collided with protons and every now and again made positrons. Because positrons want nothing more than to destroy themselves with electrons and create light, Reines and Cowan measured the resulting light signature and, in the process, proved that neutrinos are real things. What could be simpler?

Neutrinos are *so* reluctant to interact with other particles that if I were to fire a neutrino through a light-year's worth of lead, it would have about a fifty–fifty chance of making it through completely unscathed. Fortunately, we need to see only a few of them to learn an awful lot about how they work. By building giant detectors underneath mountains—oddly reminiscent of the Dwarrowdelf in Middle Earth—it's possible to detect a half dozen or so neutrinos a day.

Neutrinos do play a very important role in our lives. The three fundamental forces I mentioned earlier are the strong nuclear force, electromagnetism, and gravity. I left one out: the weak force. Whenever weak interactions occur, there's almost always a neutrino involved in some way. Weak though it may be, the weak force is what allows the sun to turn hydrogen into helium and, as a by-product, the heat and light that make life possible on earth. No weak force, no sun, no us.

For the most part, the weak force works exactly the same in our antimatter Wonderland as it does here on earth, but there is a very subtle difference, which shows up in a property known as spin. Though it may sound familiar, spin is weird, far weirder than it will appear here at first blush.

You can imagine a particle, an electron, for instance, as a little spinning orb of charge. Electron spin is different from the spin of the earth. The earth spins on its axis once a day. This is, in fact, the *definition* of a day. But here's the rub: Because of the tidal interaction with the moon, the length of a day is ever so slightly speeding up by about 2 milliseconds every century.*

For subatomic particles, you can't change the amount of spin—ever. Every particle that we've yet discovered has an intrinsic, immutable

* Superman can, of course, speed it up or slow it down more efficiently.

spin, including the neutrinos we talked about a moment ago. Neutrinos, electrons, and even protons, for that matter, cannot be slowed down or sped up under any circumstances.

For some particles, the ones with charge, measuring the direction of the spin is relatively easy. You measure the direction of the spin the same way you find the poles of earth, using magnets. The earth has molten iron in the core, and as the earth rotates, the iron generates a giant magnetic field. We can then measure this magnetic field using *another* magnet, though you probably know it as a compass.

Electrons are much the same way. As they spin, they generate small magnetic fields. Looking at the electrons from above, they can spin in one of two different ways. An electron spinning counterclockwise is described as spin-up, and a clockwise spin is called spin-down.

To figure out which end is up, we can run electrons through a device consisting of a pair of ordinary magnets and see which direction the electron gets deflected. As drawn below, the ones that get deflected upward are spin-up and the ones deflected downward are spin-down.

We could orient our magnets any way we like. The *up* and *down* in

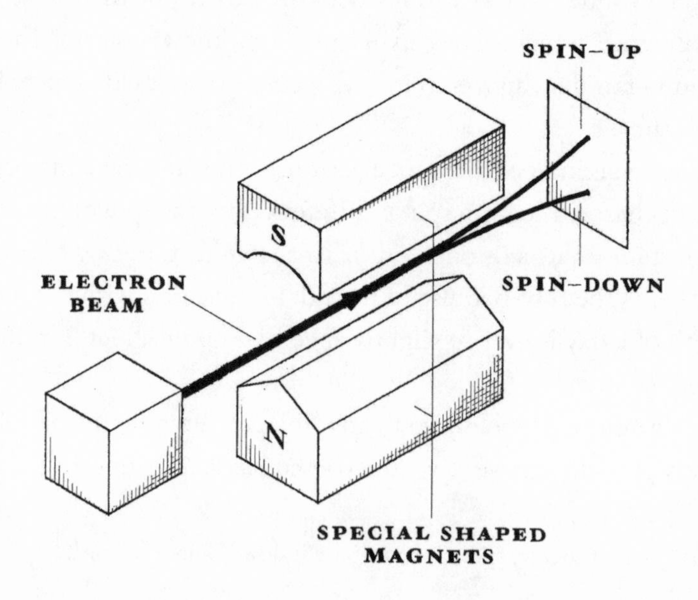

our experiment have no connection to the orientation of the solar system or anything else, but it will help our sanity tremendously if we just agree to set up our coordinates so that our respective ceilings are in the upward direction.

There's a strange, and very nonintuitive, twist to spin. If the magnets are oriented vertically, the electron will be measured to be *either* spin-up or spin-down, never somewhere in between. This is far different from the spin of the earth, which is tilted about 23½ degrees to the plane of the solar system. Likewise, if you turn your device to measure the horizontal spin of an electron, you'll find that it's *either* spin-left or spin-right. This is part of the magic of quantum mechanics.

But that's not the strangest spin fun fact. Let's say you have a nuclear decay and *poof!* out pops a neutrino. Every single neutrino will come out spinning clockwise if you're looking at it face-on. Because neutrinos are so hard to measure, we have to infer this indirectly from the spins of positrons and whatnot, but it seems to be a hard-and-fast rule of the universe.

Antimatter is exactly the opposite. Antineutrinos coming out of a nuclear decay will spin counterclockwise. Matter and antimatter seem to know the difference between left and right, which is, quite literally, the only major difference between the two. Just like you and that person in the mirror you see every morning, sort of.

It seems like a trivial difference, and one that would require tens of millions of dollars of equipment to adequately investigate, but if you're paranoid that you've fallen through the rabbit hole without anybody noticing, you do have some recourse.

Let me anticipate a collective eye roll from everyone reading this. We go through all of this buildup about the difference between matter and antimatter, and the only thing we can find is that a particle we never directly deal with spins in the opposite direction from the antiparticle? Bear with me for a bit because this little difference in spin is just the tip of the iceberg.

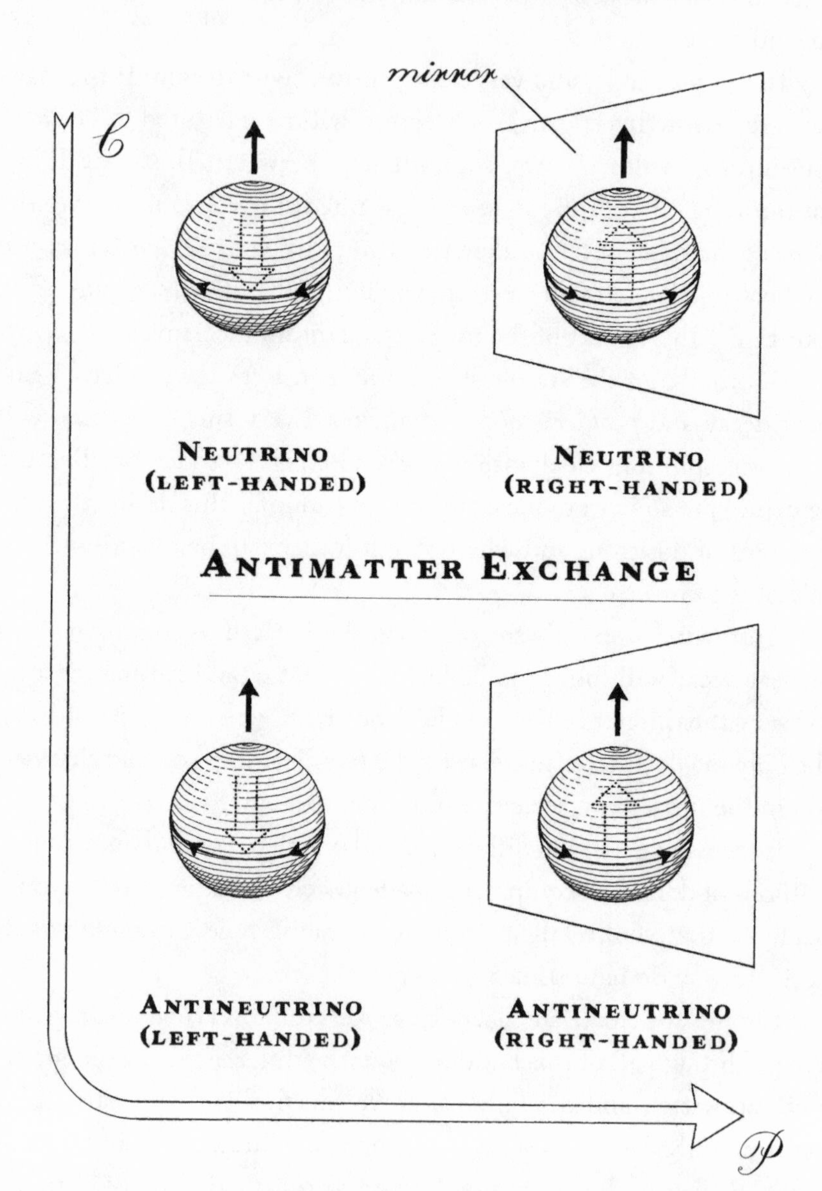
SPIN OF THE NEUTRINO & ANTINEUTRINO
mirror
C
NEUTRINO (LEFT-HANDED)
NEUTRINO (RIGHT-HANDED)
ANTIMATTER EXCHANGE
ANTINEUTRINO (LEFT-HANDED)
ANTINEUTRINO (RIGHT-HANDED)
P

PHYSICS THROUGH THE LOOKING GLASS

Matter and antimatter are *almost* identical to one another, except for the tiniest of differences: Neutrinos spin one way and antineutrinos spin the other. That's a fundamental asymmetry of the universe, but it isn't really the whole story. Let's get back to the guy in the mirror.

You could imagine that the antimatter world would look *just right* if we looked at it in a mirror. I happen to be right-handed, but my looking-glass version is apparently left-handed. Spin works the same way. A left-handed neutrino looks right-handed in the mirror.

Mirror symmetries are among the most pervasive and pleasing in all of nature. Almost all vertebrates are bilaterally symmetric at least on the outside, and we're clearly genetically conditioned to find that hot. Think of poor Narcissus, who saw his reflection in the mirror and was so struck by his own beauty that he was forever frozen in place, and ultimately turned into a flower. If humans had a clear left–right asymmetry, the figure that he saw in his reflection would have been so awkwardly unfamiliar that he would have recoiled in disgust, and this tragedy could have been averted.

This goes beyond the physical. Literary and orthographic geeks get turned on by reflection symmetries in words and phrases: palindromes, which read the same forward and backward. There's something that appeals to the cleverness of the human mind in phrases such as "Able was I ere I saw Elba" and "Go hang a salami—I'm a lasagna hog." Palindromes show up artistically—notably in the works of M. C. Escher—and even musically. In his classic *Gödel, Escher, Bach,* Douglas Hofstadter describes the *crab canon* of J. S. Bach, which plays the same forward and backward.

But most things, at least in the human world, look different from their mirror images. In most places, we read from left to right. Leonardo da Vinci (who had a great deal to say about symmetries in general) co-opted the symmetry of written language by famously making his notes left-handed and writing backward, as did Lewis Carroll, who originally introduces *Jabberwocky* by writing it in reverse.

Likewise, drivers in most countries stick to the right side of the road. But nothing prevents us from imagining a backward mirror country filled with unimaginable horrors in which they drive on the left, serve their beer warm, and da Vinci's notes look perfectly normal.

Reflection asymmetries even show up in our biology, at least internally, because our hearts are on the left-hand side of our chests. Much like cars, people look more or less symmetric on the outside, but there are asymmetries on the inside that are really just accidents of history.

Our DNA spirals a very specific way. If you point a strand toward you, it will *always* twist counterclockwise. This is a right-handed spiral. Screws work the same way; no matter how you twist and turn a screw, the threading is the same. It's always righty-tighty, lefty-loosey.*

This holds for all of the DNA of all of the creatures on earth. A biologist could most definitely tell that he or she was looking at DNA through a mirror. This single-handedness, in fact, is a very good argument for a common origin of all life on earth.

Shine a light through a solution of sugar water. Sugar, remember, is extracted from the sugarcane plant, which means that the molecules are ultimately a product of biology, not just chemistry. Naturally occurring sugar molecules fold a particular way and will cause the light to get polarized, spinning more in one direction than the other. Then just steal a set of 3-D glasses from the movie theater. One eye will allow you to see only left-handed polarized light and one eye will let you see only right-handed polarized light. Seen through the sugar water, the right-handed lens will appear brighter than the left-handed one.

How do sugars know the difference between left and right? The molecules themselves, as in DNA, spiral in a particular direction, one

* They do make screws with left-handed threads, but they're the exception.

that looks exactly opposite in the mirror. A left-polarizing sugar would be chemically identical to a right-polarizing one, but were we to fill a dish with bacteria and artificially create left-polarizing sugar (the mirror image of our "real" sugar), the bacteria would go hungry, unable to eat any of it. The enzymes used to digest the sugars are themselves asymmetric, suited to work only with left-handed sugar. Why, after all, would they evolve any other way? Or, as Alice says in *Through the Looking-Glass,* "Looking-glass milk isn't good to drink."

We have to take a good hard look in the mirror to understand why—or even if—there's any *fundamental* difference between left and right. It's easy to imagine a planet that has exactly the same laws of physics, yet people have their hearts on the right, have a written language that's an exact reverse of ours, and so on.

These sorts of asymmetries are important for us to keep in mind. They're not hardwired in; they just happen. But once they happen, it becomes very, very difficult to switch. Take a drive on the left-hand side of the road and see how that works out. But in the universe in the rear-view mirror, *everyone* drives on the left.

Like antimatter, the mirror universe is not so different as you might suppose. Richard Feynman illustrates the point effectively:

> Suppose we build a piece of equipment, let us say a clock, with lots of wheels and hands and numbers; it ticks, it works, and it has things wound up inside. We look at the clock in the mirror. How it *looks* in the mirror is not the question. But let us actually *build* another clock which is exactly the same as the first clock looks in the mirror—every time there is a screw with a right hand thread in one, we use a screw with a left-hand thread in the corresponding place on the other. . . . If the two clocks are started in the same conditions, the springs wound to corresponding tightnesses, will the two clocks tick and go round, forever after, as exact mirror images?

Our intuition, and virtually any experiment that you're likely to be able to do in your basement lab, suggests that if we look at the reversed clock in the mirror, then it should look identical to and run the same way as the original.

Suppose Alice found herself through the looking glass, in a second parallel world in which everything was a reflection of earth. Would she be able to tell? To put it another way, would she be able to figure out which hand is *really* her left?

This is a trickier one than the antimatter switch because it's almost impossible to put yourself in the situation completely. Your gut reaction might be something along the lines of "Of course she'd notice. Don't ask stupid questions."

But remember when you were younger, and you occasionally got your left and right confused. How did you remind yourself? You'd hold up the thumb and forefinger on your hands, and the one that made an L would be your left hand.

That's not going to work, though. The looking glass also reflected all of the letters, so that now L is backward. Alice would identify her right hand as the left one. She's not going to figure out whether she's in mirror-earth or the ordinary one by simply looking at her hands.

This shouldn't come as much of a surprise. If things *didn't* look reasonable in the mirror, then I wouldn't constantly get fooled into thinking there's another room in restaurants with wall mirrors. Every time.

I want you to keep something in mind. It's not that the antimatter Wonderland or the looking-glass universe are identical to our own. They clearly aren't. But the question, as with Feynman's clock example, is whether the rules in those universes are identical to the ones in ours or whether they differ in some subtle way.

Alice can jump up and down, play with magnets, measure the detailed structure of an atom. All of these things will produce the exact same results as before she went through the looking glass. If everything works exactly the same in the real world as it does in the mirror (it doesn't), then we'd have a symmetry of nature:

> **P Symmetry:** All of the laws of physics look just as valid if you view everything through a mirror.

The *P* stands for "parity." We already *know* that this isn't a perfect symmetry for our universe. When a particle, an electron or a neutrino, for example, gets created in a weak reaction, it is *always* left-handed, which is just another way of saying that if it's headed toward you, it'll look like it's spinning clockwise. Antiparticles, on the other hand, have exactly the opposite spin. That's it.

That's the difference. C and P Symmetries aren't the same, but they are *very* intimately related. Neutrinos and antineutrinos differ in exactly two ways: They are antiparticles of one another (C), and they have

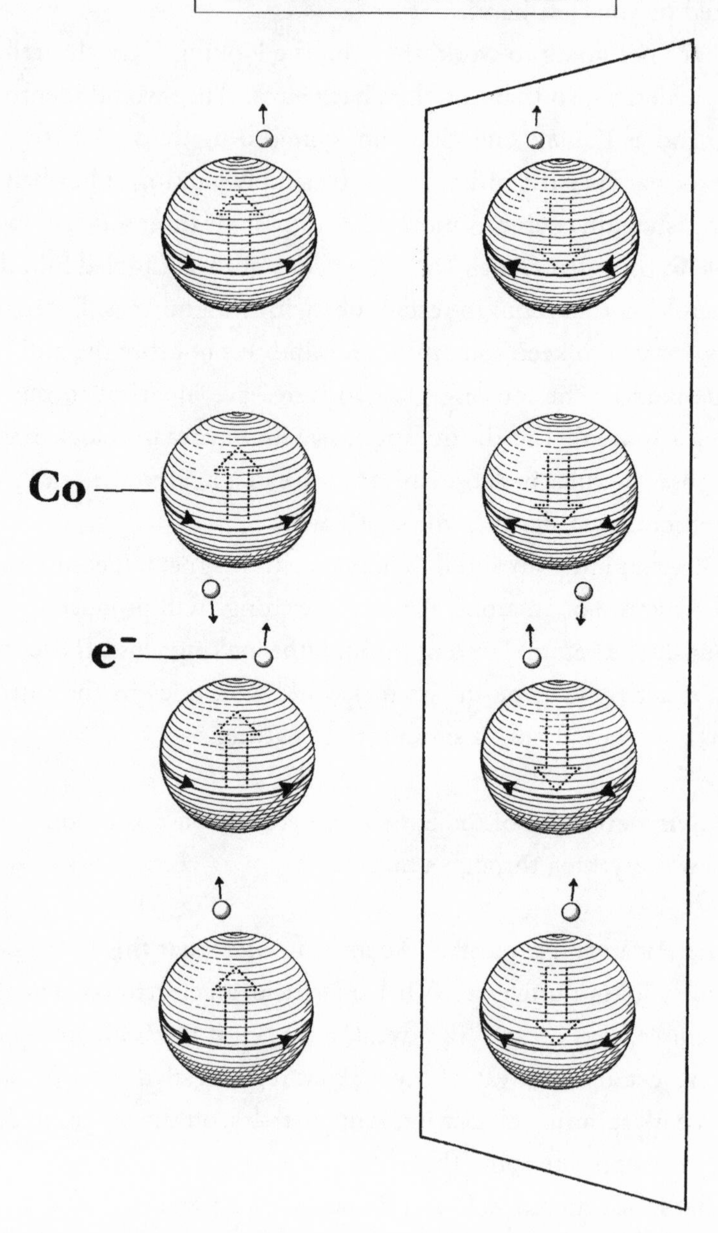
COBALT-60 DECAY
Co
e⁻

opposite spin (P). Neither is a perfect symmetry of physics on its own, but the combination of the two seems *almost* like a symmetry of nature. Take the antimatter version of a left-handed neutrino and look at it in a mirror and you get a right-handed antineutrino. The starting and end states are different, but both the left-handed neutrino and right-handed antineutrino are things that exist.

Alice doesn't need to look at particles as elusive as neutrinos to realize that things are going to be a bit different through the looking glass. In 1956, C. S. Yu and her collaborators performed an experiment with a radioactive isotope of cobalt. They set the spin of the cobalt atoms in a particular direction. Imagine that if you looked at the atoms from above, they all appeared to be spinning counterclockwise, spin-up.

When the cobalt decayed, electrons were produced. Surprisingly, most of the electrons flew upward. Sounds like a simple result: Cobalt decays into electrons going in the same direction as the spin.

Oh, you're not surprised? You should be.

To realize how strange this result is, you have to think about the whole thing in a mirror. Mirrors change the spin of particles from one direction to another. In the mirror, the cobalt atoms spin clockwise, which means that they spin downward. The electrons, on the other hand, are still emitted upward, whether you look at them in the mirror or otherwise.

The electrons' emission and spin line up in our universe but are opposite in a mirror. Finally, a concrete experiment that you can do to see if you've fallen through the looking glass.

MIRRORS AND ANTIMATTER

All of this mucking about with mirrors and whatnot may have obscured an important question that seems to have fallen by the wayside. What was the origin of all matter in the universe again? Oh yes. That detail.

To figure it out, we're going to need to imagine yet *another* parallel universe:

1. Take every particle in the universe and turn it into its antiparticle (and turn every antiparticle into the regular particle).
2. Look at the whole thing in the mirror.

The $64,000 question is, Does the new universe, a looking glass/ Wonderland combo, have the exact same laws of physics as we have in the "real"* one? The combination of the two is known as a CP transformation.

Imagine you have an electric current running through a wire. An electron has a negative charge, and a proton has a positive one. As electrons move through a wire, the current flows in the *opposite* direction. Say your electrons are flowing to the left, the current is then going to the right. Take the Wonderland (antimatter) version, and now *positrons*

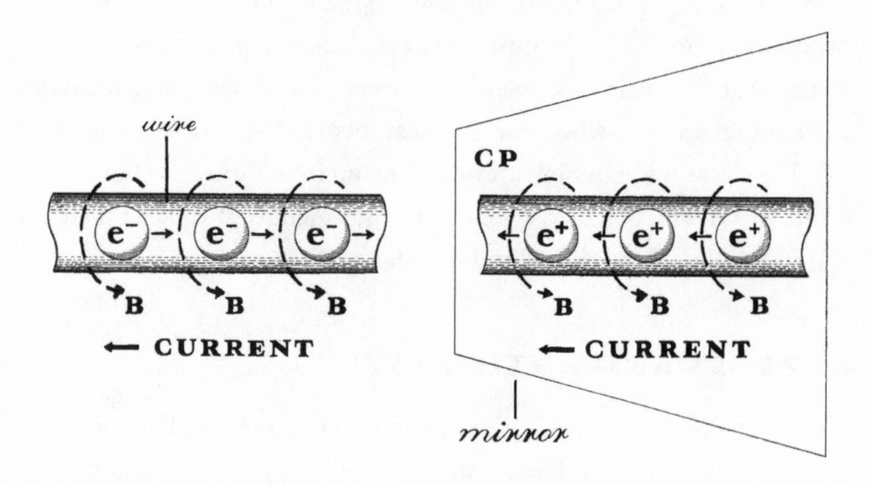

* Scare quotes inserted to make you question the legitimacy of your existence.

are flowing to the left. Flip the wire in a mirror and now the *positrons* are flowing to the right, producing the exact same current as the original. This is kind of a big deal because currents give rise to magnetic fields, and so under a CP transformation, a wire produces the exact same magnetic field as the original.

Electromagnetism passes the test, but not every experiment is quite so obliging.

In 1967, the Soviet physicist Andrei Sakharov found the minimum conditions necessary to get around the matter–antimatter asymmetry problem, and in short it requires that *something* needs to change if you perform a CP transformation on the universe. As the pros put it, there needs to be a violation of CP Symmetry.

Your existence—and the general dominance of matter over antimatter—is a pretty strong argument *against* a perfect CP Symmetry, but experimental evidence, or at least the experiments we've seen so far, would seem to suggest otherwise.

We've already seen that you can learn a lot about how the universe works by looking at how particles decay. At very high energies, particles known as kaons can be produced in particle accelerators, along with their antiparticles. If you've never heard of kaons before, don't feel too bad. They live, on average, for only a few billionths of a second before they decay into lighter particles, which themselves typically decay very, very quickly. You're unlikely to encounter a kaon during the normal course of your day.*

That's okay, because the real action happens once kaons fall apart. In 1964, James Cronin and Val Fitch, both at Princeton University, were doing postmortem analysis on kaons and they found something unexpected. Kaons and antikaons, particles that everyone assumed

* Incidentally, if you're finding all of these new particles a bit overwhelming, don't sweat it. For one thing, we've seen almost everything we're going to see. And for another, there's a handy cheat sheet at the end of the book.

were identical, seemed to decay differently from one another.* There was a distinction between matter and antimatter.

It's even more insidious and subtle than you might first suppose. Kaons and antikaons slowly oscillate back and forth between one form and the other, like day into night and back again.

On average, day and night will last the same duration, but clearly that symmetry is sometimes broken. During the summer, for instance, the day lasts longer than the night. By the same token, the symmetry between matter and antimatter would suggest that the particles *should* spend half their time as kaons and half their time as antikaons, and even though we don't know which it is ahead of time, we *can* tell which state and what type of particle it was before it decays.

If you start with a kaon, every so often it'll decay into an electron and some additional detritus that we're not going to worry about. On the other hand, if you start with an antikaon, every so often, it'll decay into a positron and some still-different detritus.

The reasoning goes that if you start, instead, with a giant pile of kaons and antikaons, the two should oscillate back and forth, and so on average you'd expect—in a universe with perfect CP Symmetry—an equal number of electrons and positrons to come out.

This is *not* what happens.

Instead, such experiments produce a slightly larger number of positrons than electrons. I don't want you getting too hung up on whether there are more positrons or electrons. The main point here is that you can't simply swap all of the matter for antimatter in the universe, even if you subsequently look at everything in a mirror, and have things look the same. The combination of charge and parity symmetry simply *does not hold* in our universe. This was a very big deal, for which Cronin and Fitch were awarded the Nobel Prize in 1980.

* A kaon koan: What is the sound of a subatomic particle turning into its antiparticle?

Since Cronin and Fitch's experiments, there have been a number of similar and even more dramatic results, all of them suggesting something similar: There is something asymmetric between matter and antimatter that seems to show up when the weak force is involved. You should know, though, that none of these experiments actually resulted in an excess of matter being created over antimatter, just that matter and antimatter decayed differently from one another.

But all of this doesn't ultimately tell us *why* there is a difference between matter and antimatter. What were the reactions that allowed the creation of more of one than the other? That, after all, is the ultimate answer to the question of where we came from.

No one has yet figured out exactly how things played out in the early moments of the universe. All we know is that we exist because of some sort of symmetry violation in the universe from *very* near the beginning. The conditions in the early universe were *extremely* hot—maybe that had something to do with it.

Every now and again, you'll hear accelerators described as "recreating the conditions of the Big Bang." This is more or less right. The universe was hotter, and more energetic, in the past. The closer to the Big Bang that you want to explore, the hotter it is. Nothing we've seen so far in particle accelerators has given even the slightest inkling of producing a net matter over antimatter. The current thinking is that the small matter–antimatter accounting error occurred very, very early on, around 10^{-35} second after the Big Bang during which the temperatures were more than a quintillion times those at the center of the sun. Suffice it to say, we're not able to produce those energies in a lab.

Even at those astounding energies, the asymmetry between matter and antimatter is extremely small. For every billion antiparticles that were created, there were a billion and one particles. One. Just one. We know that because there are currently about a billion times as many photons in the universe as there are protons. When the billion antiprotons annihilated with the billion protons, they left behind the billions of photons that we observe today, though greatly weakened by the expansion of the universe.

Eventually, all of the antiparticles annihilated with almost all of the particles, leaving the one part in a billion to make all of the "stuff" that we now see. As Einstein put it:

> I used to wonder how it comes about that the electron is negative. Negative-positive—these are perfectly symmetric in physics. There is no reason whatever to prefer one to the other. Then why is the electron negative? I thought about this for a long time and at last all I could think was "It won the fight!"

To put it another way, you're essentially a rounding error from around 10^{-35} second after the Big Bang. Doesn't make you feel very important, does it?

Of course that's just as much a bummer for the anti-people too.

Chapter 2

ENTROPY

IN WHICH WE EXPLORE WHERE TIME COMES FROM OR WHETHER IT JUST *IS*

I don't think I'm alone here in envisioning a glorious future where we'll all be jetting through the universe in galaxy-class starships. Hell, one of my main motivations in writing this book is the faint hope that one of you might decide to take things a step further and actually figure out how to build a warp drive. But before you start bending what we know of the laws of physics, it's incumbent on me to issue a few words of warning about what awaits. I'm not talking about exploding stars or Vogons (but those too). I'm talking about losing your way.

On earth, we have all sorts of handy reminders to help us find our way: gravity, the North Star, the magnetic pole of the earth. But in deep space, there is no up or down, left or right, north or south.

You might comfort yourself with the thought that while we may get lost in the three dimensions of space, we should at least be safe in time. Time *feels* solid, immutable, and real. Left and right may be more or less interchangeable, but the future and the past are *very* different. Right?

Objects in mirrors generally look perfectly ordinary, but intuitively, a "time mirror" seems absurd. Run the universe, or even your day, backward in time, and events will unfold in a decidedly different manner than if you played it forward. If you've seen *Memento* and could figure out the chain of events the first time through, I congratulate you. Now imagine living your life in reverse.

There's a little thing called causality, for instance. You do a thing and another thing happens because of it. Flip the clock of the universe, and all of a sudden effects start happening before their causes, and all hell breaks loose.

Stupid physicists! Why even bother talking about the arrow of time when it so clearly has a preferred direction?

Easy there, champ. The arrow of time is far more reversible than you might first suppose.

HOW SPACE AND TIME ARE THE SAME . . . OR NOT

Life is a journey. There's the literal sense of journey where you actually move about in space and see things, but you're traveling around in time as well. It's just that you're moving along at 1 second per second, and that sort of motion seems like the most natural thing in the world. While time *is* different, time and space are far more alike than we normally think.

Just as the speed of light is the currency conversion rate between matter and energy ($E = mc^2$), it's also the rate of conversion between space and time. You've probably heard of a light-year; it's simply the distance that light can travel in a year, about 10^{16} meters. If you don't have a handle on that (and who could blame you?), it's about a quarter of a way to the nearest star, Proxima Centauri.

Or if you want to think about it in terms of human engineering limits, the *Voyager 1* spacecraft was launched by NASA back in 1977 and has been flying out of our solar system since. *Voyager* is the most distant man-made object ever launched, now nearly 20 *billion* kilometers from earth. Light could cover the distance in a bit under 17 *hours*.

There's a close relationship between space and time. Many physicists treat them the exact same way, talking about seconds and light-seconds interchangeably and carelessly setting the speed of light equal to one. Practically speaking, that's how we now *define* units of distance and time.*

In 1983, the very grand sounding seventeenth General Conference on Weights and Measures *defined* the second in terms of the "hyperfine transition" of cesium-133. Every now and again, a cesium atom will give off light, so the conference defined a second as 9,192,631,770 times the period of the emitted photon.

Once you know what a second is, figuring out distance is a piece of cake. A meter is simply defined as the distance that light travels in 1/299,792,458 second.

The consequence of the finite speed of light is that we're always looking into the past. The sun we're *seeing* now isn't the sun as it *is* right now. It's the sun of 8 minutes ago. It could have burned out 7 minutes ago, and we'd have no way of knowing. When Neil Armstrong gave his line about "one small step . . ." we were listening in on history, both literally and metaphorically because the radio waves to transmit his message took about 1.3 seconds to reach us.

Even in rather mundane cases like reading a book, you're looking back in time. Hold the book about 1 foot away from your eyes and you're peering 1 billionth of a second into the past.

Although moving around in space and time are, in some sense, the same, I want to focus on the differences. For one thing, you're moving through time *much* more quickly than you're moving through space. In

---

* Of course, you can take the time–space relation too far. Boasting that your ship can make the Kessel Run in under 12 parsecs, for instance, makes you come across as a jackass.

1 second, you travel through *1 second* of time (of course). But even the very fastest man-made satellites cover only about 0.2 *milli*-light-second of space in 1 second of time. They may as well be standing still.

We travel through time so much more quickly than we do through space as a natural consequence of the insanely fast speed of light. Light is so fast that up until the last few centuries, we couldn't really be sure that it didn't travel infinitely quickly. In a very real sense, to understand time, we really need to start by understanding space.

The first measurements of the distance to the sun, for example, were done entirely from geometry, and the distance to the sun, now known—a distance of about 93 million miles, and referred to, a tad unimaginatively, as an Astronomical Unit—was first measured in multiples of the earth's radius.

The ancients, the ones who weren't foolish enough to think that the sun orbited around the earth, tried a number of relatively unsuccessful approaches to figure out this fundamental step in the distance ladder. We don't actually know *how* unsuccessful they were, because it's tough to figure out exactly how to convert between ancient units. Working in the third century BCE, Aristarchus of Samos, in one of the best estimates until modern times, was off by something like a factor of 15.

It wasn't until nearly 2 millennia later, in the latter 1700s, that the French astronomer Jérôme Lalande used a rare (and highly anticipated) astronomical event to accurately measure the distance to the sun: the transit of Venus.

Every century or so, the planets line up just perfectly so that the planet Venus passes precisely between the earth and the sun. Transits are useful because they appear slightly different to people on different parts of the earth. Two observers aligned east–west from one another will see the transit start at very slightly different times.

It works exactly the same as your eyes.* Your left and right eyes see

* With apologies to cyclopes for my insensitivity.

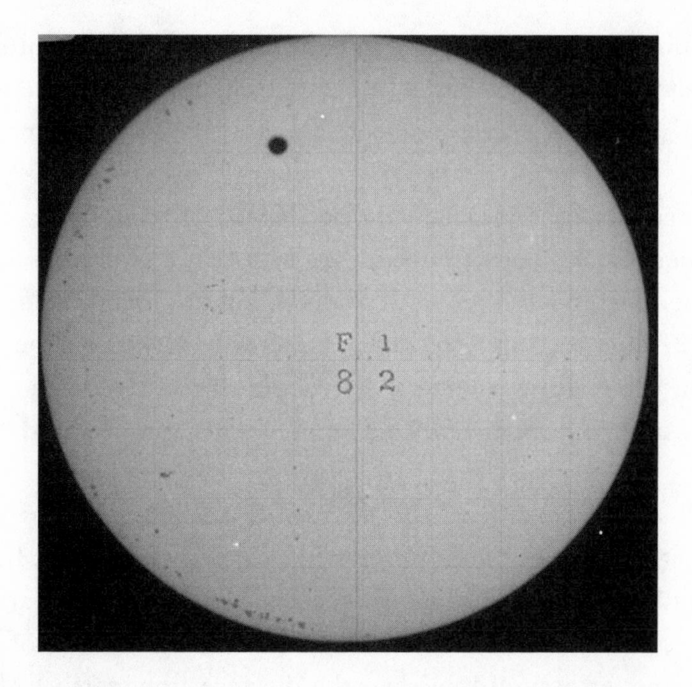

ever so slightly different images, and from them, your brain does a calculation of depth and distance. Blink your eyes back and forth, and you're going to see things move slightly, with closer objects moving more. Put into mathematical terms, your brain is really figuring out all distances as a ratio of the distance to the separation of your eyeballs.

Because of relative tilts of earth's and Venus's orbits, you get one shot to observe a transit of Venus, and then must wait 8 years for another. After that, you're out of luck for about 120 years. The last transit of Venus was seen on June 5–6, 2012, so if you missed it, you've probably missed it for your lifetime.

Lalande had the good fortune to be in his prime during the 1761 and 1769 transit. While he didn't take the data himself, he had access to the measurements of others and was able to get a *very* good measure of the distance to the sun, accurate to within a few percent.

Although it's true that we weren't able to measure the distance to the sun in meters until the 1770s, as it turns out, we knew the approximate distance to the sun in light-minutes almost a century before that. Back in the 1670s, a Danish astronomer named Ole Rømer noticed something very odd about the recently discovered moons of Jupiter.

It may not have escaped your notice that orbiting bodies form a convenient clock.* For example, of the four bright Galilean satellites, the innermost moon is called Io,† which has an orbital period of 42 hours, 27 minutes, and 33 seconds. Rømer was able to note the phases of the moons relative to Jupiter when the planet was at opposition (closest approach). Then, roughly six months later, he observed Jupiter again.

He should have been able to predict the exact phases of the moons at any time he liked. They ran like clockwork, after all. Instead, Rømer found that when Jupiter was nearest in its orbit to the earth, the moons seemed to be running about 22 minutes ahead of when Jupiter was farthest away. When the distance from earth to Jupiter was a maximum, Rømer might have expected the innermost moon, Io, to pass in front of Jupiter at 9:00 (calibrated from when the earth and Jupiter were nearest to one another), only to have to wait until 9:22.

Rømer concluded (correctly, as it happens) that light takes a certain amount of time to get to us from Jupiter, and when it's farther away, it will take longer than when it's closer. Because at closest approach Jupiter is 2 Astronomical Units closer than at farthest approach, Rømer found that it takes light about 11 minutes to travel an Astronomical Unit.

---

* If it has, let me point out the similarity of the words *month* and *moon* and let you work out the rest of the relation for yourself.

† Named, like a lot of Jupiter's moons, for one of Zeus's many, many lovers. Io was a nymph. Just for funsies, I'd recommend you look up the mythologies relating to the other Galilean satellites.

Astronomical measurements are tough, and were even tougher in the seventeenth century because telescopes were still in their infancy. As it turns out, the actual light travel time to the sun is closer to 8 minutes, 19 seconds. Still, Rømer was in the right ballpark.

HOW MUCH HISTORY *IS* THERE?

As I said, while light describes how time and space are similar, there are clearly ways in which the two are different. There's no special direction in space, but there's clearly a preferred direction in time. There's a difference between the past and the future. Most significant, the single most important event in the universe—its beginning and hence the beginning of time itself—is in the past. That is, of course, if the universe actually *had* a beginning.

Everyone has heard of the Big Bang. But it might not be immediately obvious *why* there needed to be a Big Bang.

When Einstein first proposed his Theory of General Relativity in 1915, he assumed that the universe was eternal, and he even jury-rigged his equations to make sure that it was. Without his tweak—adding a value to the field equations, known as the cosmological constant—a universe either expands forever or expands for a while and ultimately collapses.* Einstein added his constant to precisely counter the gravitational attraction of matter in the universe and keep everything static. He was, as biographer Walter Isaacson pointed out, almost immediately apologetic about it. In Einstein's own words:

> We admittedly had to introduce an extension of the field equations that is not justified by our actual knowledge of gravitation.

---

* At which point presumably the heavenly angels would play the universe's saddest trombone.

That extension has been called a fudge factor. Before you call out intellectual hackery on Einstein, consider the question of eternal time from the other direction: why it's weird to live in a universe with a Big Bang. Once you acknowledge a definite beginning to the universe, then you immediately start asking questions about why we're alive *now* rather than a billion years earlier or a trillion years later.

We couldn't live at any time in the past and future history of the universe. The earth has been and will be hospitable to our existence for a surprisingly short time, for instance. Supposing we don't find some other way to kill ourselves first, in about 4 billion years the sun is going to turn into a red giant, frying everything on earth that managed to survive that long. But this is just an eyeblink by cosmic scales. Although the universe has been around for only 14 billion years so far, as near as we can currently tell, it will continue expanding literally forever.

The existence of complex life presupposes certain conditions favorable to life. You need a certain amount of energy to sustain any activity, life or otherwise, heavy elements are normally required to carry on any complex chemistry, and so on. This is known as the Anthropic Principle.*

The universe varies from place to place and changes over time. All that the weakest versions of the Anthropic Principle claim is that people are likely to find themselves in those regions of space and time most suited to their evolution and existence. In other words, we're here because if we weren't, then we wouldn't be here to ask the question of how lucky we'd have to be to live in one of the few places conducive to life.

See how circular the argument is?

In the potentially quadrillions of years that the universe has left in it, stars like our sun will last for only a tiny fraction of it. In a global

* Despite the "anthro" prefix, we're not necessarily talking about talking, hairless apes.

sense, we're living at the dawn of the cosmos because most of what remains will be dark, cold, and incredibly inhospitable. We're living 10 billion years or so after the Big Bang because, as near as we can tell, it's more or less the only time we *could* live.

It's easy to forget the general hostility of the universe to life and suppose that all planets and epochs of universal history are like our own. Movie aliens reflect at least some of our assumptions about what's out there. Even the ones that aren't roughly six feet tall and vaguely humanoid still have a bilateral symmetry, humanlike needs and wants, and (despite additional millions or billions of years of evolution) more or less human intelligence.

Charles Darwin's work in the nineteenth century made a very compelling case that people hadn't been around forever, and shortly after his work, geological evidence suggested that the earth was of a finite age as well. Einstein was aware of all of this, but there was a *lot* we didn't know about the universe as a whole in the early twentieth century.

It wasn't until 1920 that Sir Arthur Eddington (the guy who thought only he and Einstein understood relativity) discovered that it was nuclear fusion that powered the sun and other stars. His insight would, of course, have been impossible without Einstein's discovery of the equivalence of matter and energy. Without those key pieces of evidence, we had no real way of figuring out how old the sun was, let alone the universe.

In 1924, 9 years after Einstein formulated General Relativity, Edwin Hubble discovered that the Milky Way wasn't the only galaxy (or *nebula*, as he called it) in the universe. It was another 5 years after *that* that Hubble found that nearly all of the galaxies in the universe are receding from us and that the more distant galaxies are receding faster than are the nearby ones. This was the smoking gun of an expanding universe—one that had a beginning.

Einstein's version of the steady-state universe wouldn't have worked in any case. Even if the cosmological constant were just perfectly tuned to keep the universe as a whole from expanding or collapsing, little patches of the universe would still collapse on themselves. His whole model was incredibly unstable.

Rather than be embarrassed by Hubble's discovery, Einstein was thrilled with the result:

> The people at the Mt. Wilson observatory are truly outstanding. They have recently found that the spiral nebulae are distributed approximately uniformly in space, and they show a strong Doppler effect, proportional to their distances, that one can readily deduce from general relativity theory without the "cosmological" term.

Not everyone was quite so sanguine about the idea of a beginning for the universe. Fred Hoyle (who coined the phrase *Big Bang* to *make fun* of the theory) and others tried to get rid of the Big Bang model entirely by imagining a steady-state model in which the universe was expanding but matter was continuously being created to fill in the gaps. It's not as ridiculous as you might first suppose.

We tend to forget how empty the universe is on average, which means that in Hoyle's steady-state model, only the tiniest amount of energy would need to be created. In the entire lifetime of our sun, a volume of space the size of the earth would need to generate only a couple of milligrams of matter. Trust me; you wouldn't even notice.

The problem with Hoyle's steady-state universe is that there's no reason to suppose that matter really *is* continuously created. What's more, we're now able to look at the history of the universe (that is, at objects at many different distances from the earth), and it is pretty clear that the universe, overall, is changing.

In the last couple of decades, observations of distant explosions known as supernovas have demonstrated that the universe is not only expanding but accelerating, exactly what we might expect if the universe had some sort of large cosmological constant. We tend to call it *Dark Energy* today, but the idea is much the same: There is some unchanging fluid throughout the universe that seems to work as a sort of antigravity. This was of such staggering importance that the leaders of the collaborations, Saul Perlmutter, Brian Schmidt, and Adam Riess, were awarded the 2011 Nobel Prize in Physics.

The accelerating universe has some interesting consequences. The universe will continue to expand forever, and as it does, it will get colder and colder, essentially without limit. The stars will burn out. Some will turn into black holes, and even those will eventually evaporate. Protons (in principle) will decay into radiation, and the radiation throughout the universe will become colder and colder and more diffuse.

The future is bleak; the past is a nightmarish hellscape of fire and chaos. In a very real sense, we are living in a particularly auspicious moment in history.

But just because the universe can tell the future from the past doesn't mean that the laws of physics can. Can they?

OKLO

One of our most fundamental assumptions is that the laws of physics don't change with time, even if the apparent effects of those laws *do* change. But you know what happens when you assume, don't you?

We always need to entertain the possibility that the laws change but have somehow arranged themselves in a way that's perfectly designed to fool you into thinking that they're constant. It's kind of like someone planting dinosaur bones in your backyard to trick you into thinking that dinosaurs once roamed the earth. It's all so implausible.

We live at *this* time, and any experiment that human beings could perform could last at most only a few hundred years out of the total 13.8 billion years of the universe. Even if the laws of physics *were* changing, it's doubtful they'd change quickly enough to be measured directly by us puny humans.

This means that statements about the unchanging nature of physical laws necessarily need to use those same physical laws to make models about the past. The upshot, though, is that models from the past produce a staggeringly consistent picture. For my part, I'm actually satisfied by this sort of argument. It is *very* hard to make completely self-consistent assumptions that end up producing self-consistent results unless those assumptions are correct.

Hard, but maybe not impossible.

There are a few natural experiments that test the timelessness of the laws of nature in a fairly unambiguous way. One of the most amazing was discovered in 1971 in the village of Oklo in Gabon. For some time previously, the French had been mining uranium in Gabon when geologists discovered what could only be described as an ancient nuclear reactor.

When I say "ancient," we're not talking about merely precivilization, we're talking 2 billion years old, significant compared to the age of both the planet and the universe. Just to anticipate the obvious: I'm not advocating some sort of *Chariots of the Gods* scenario here. It was just a small region where the leeching of minerals from ancient rocks, the flow of rivers, and some hungry, hungry bacteria conspired to deposit a far higher concentration of uranium than is typical anywhere outside a nuclear reactor.

Though it's all fairly nasty stuff, there are many different isotopes of uranium, and they don't all behave the same. The most common isotope is uranium-238, but the isotope that does the heavy lifting in nuclear fission is uranium-235 (U-235). For nuclear fission reactors to work, one of the crucial steps is enriching the uranium with a series of centrifuges, until U-235 makes up a few percent. In natural samples, U-235 makes up only about 0.7 percent of the total uranium on earth, but 2 billion years ago, it made up a much healthier 3.7 percent, the levels used in modern light water reactors. The ratios have changed because uranium is in a constant state of decay, and U-235 decays six times faster than U-238.

This combination of natural enrichment and a big pile of uranium produced a critical mass. The uranium fissed into isotopes of palladium and iodine, as well as a lot of energy, perpetuating the process. The Oklo site was a natural nuclear reactor that burned for millions of years.

It's amazing that such a site exists at all. But more amazing is that the detritus left behind from the nuclear fission is in exactly the proportions that we would find in a nuclear reactor today. Nuclear reactions are a tricky business. If the nuclear forces had changed over that time—and remember, we're talking about 2 billion years here—we'd be able to see it more or less directly.

Constant physical laws are more than just a curiosity, and they're far more than handy tools that allow us to say meaningful things about the early universe—though, of course, that *is* kind of a big deal. In fact, it's another symmetry:

> **Time Translation Symmetry:** All of the laws of physics behave the same at different times.

Time Translation Symmetry may seem like just a curiosity, but it turns out this is another way of saying that energy can't be created or destroyed. So this hiding-in-plain-sight symmetry is actually a *very* big deal: It allows us to do science. Any universe in which the laws could change willy-nilly would leave us completely unable to anticipate the future.

THE ARROW OF TIME

The universe appears to change over time, even though the laws of physics stay the same. Could the same sort of thing be true about the arrow of time?

In books, Merlyn from T. H. White's *The Once and Future King* and the White Queen in *Through the Looking-Glass* live their lives backward. The movie *Memento* is told in reverse to make us as disoriented as the main character, Leonard, who has lost his long-term memory. This literary shorthand tips us that different laws apply to them than to the rest of us.

But take yourself (or your favorite magician) out of the equation for a bit, and see how things play out at the fundamental level. If you look at almost all of the laws of physics, the flow of time seems to be an afterthought. Make a movie of two electrons colliding into one another and then run that movie in reverse, and the time-reversed version will

COLLIDING ELECTRONS

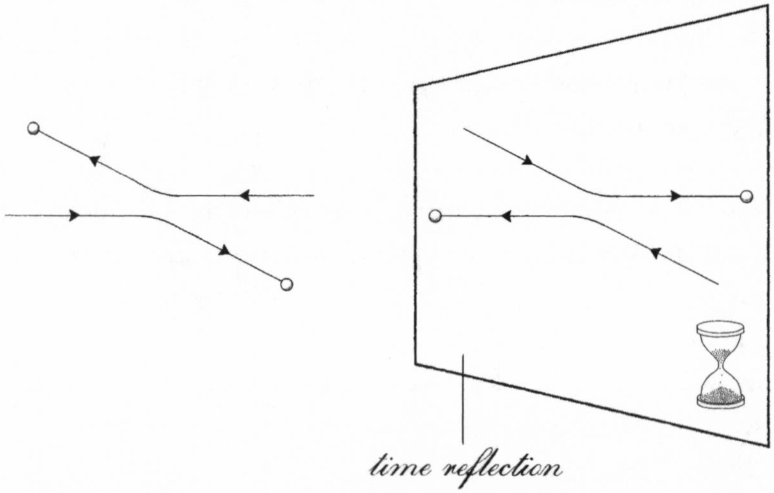

time reflection

look as normal and physically valid as does the forward version. At the microscopic level, time seems to be completely symmetric.*

We don't even need to restrict ourselves to the microscopic scale. I'll assume that at some point in your life, you've played a game of catch. The ball flies through the air in an arc, a pattern known as a parabola.† Record this and play it in reverse, and although it won't look exactly the same—it will fly from left to right, for instance, rather than from right to left—in this narrow frame, it will look equally valid.

This is yet another symmetry, and like the others, it has its own name:

> **T Symmetry:** The laws of physics look the same upon a reversal of the flow of time.

We must pause here for a moment. Look at the universe in a mirror and everything looks more or less correct. Texts are reversed, people drive on the wrong side of the road, and your heart is on the wrong side of your chest, but otherwise, you probably feel like you could adapt in short order. Likewise, switch all of the particles to antiparticles, and everything looks just fine. Of course, you wouldn't want to actually visit such a place for fear of instant annihilation.

But in a time-reversed universe, everything is going to look crazy! It certainly *seems* obvious that this can't be a symmetry of our universe. It does *seem* that way, doesn't it?

On the other hand, our intuition can often lead us astray. When we talk about the symmetries of the universe, we're not *really* talking about transforming the entire universe so much as we are about

---

* Attention, nitpickers: The exception is the weak force, but that doesn't matter for now. Don't worry, we'll get to it.

† This works especially well if you're playing catch on the moon, where there's no air resistance.

transforming the rules of the universe. Look only at a single game of catch or a single pair of electrons bouncing off one another, and reverse *that*.

So, does physics work in Merlyn's universe? Or, to put it in more respectable terms, are the laws of physics unchanged under a T Symmetry transformation? To answer this question, we need to figure out how T Symmetry relates to the charge and good old-fashioned mirror symmetries, C and P, that we've seen before. From a mathematical perspective, there is very little difference between an electron going forward in time and a positron going backward.

In his Nobel lecture, Richard Feynman related a conversation he once had with his thesis adviser, John Wheeler:

> I received a telephone call one day at the graduate college at Princeton from Professor Wheeler, in which he said, "Feynman, I know why all electrons have the same charge and the same mass." "Why?" "Because, they are all the same electron!" . . . "But, Professor," I said, "there aren't as many positrons as electrons." "Well, maybe they are hidden in the protons or something," he said. I did not take the idea that all the electrons were the same one from him as seriously as I took the observation that positrons could simply be represented as electrons going from the future to the past in a back section of their world lines. That, I stole!

Remember the example of a current in a wire, from before? If you run a movie of a positron in reverse, it produces exactly the same magnetic field as an electron running forward. We can even imagine creating a parallel universe using the following steps:

1. Turn all particles to antiparticles and vice versa.
2. Flip everything in a mirror.
3. Reverse the flow of time.

This is known as a CPT transformation, and there's almost nothing we can say about it except that literally every experiment that has ever been done has suggested that the universe is CPT symmetric. This is a big deal because, if nothing else, we've *finally* found an absolute symmetry in the universe.

Put another way, it seems as though the particle physics on a microscopic level is more or less the same whether you run things forward or in reverse. From the perspective of fundamental physics, there's nothing more special about the arrow of time than there is about which particle is the electron and which one is the positron.

And yet time definitely *feels* different from the other dimensions. But why? Short answer: We really don't know, beyond just "that's how time works." Then there's the long answer. It's the interesting one.

THE SECOND LAW

You remember the past, but not the future. You can't unscramble an egg or unmix a cocktail or unbreak a bunch of pool balls.

There's a general tendency for things to get more and more disordered. You may know this as the Second Law of Thermodynamics. The Second Law could be described rather glibly as "things fall apart," but it's really even simpler than that.

The establishment of thermodynamics is, at least in part, a direct result of the Industrial Revolution. In the 1820s, the French engineer Nicolas Léonard Sadi Carnot was simply trying to build a better steam engine and realized that, try as he might, some of the energy was always lost uselessly to heat. By 1850, Rudolf Clausius had come up with a more formal statement of what's come to be known as the Second Law:

> No process is possible whose sole result is the transfer of heat from a body of lower temperature to a body of higher temperature.

Left to its own devices, any system, including the universe, will eventually equilibrate to a uniform temperature. Everything will become uniformly disordered. Uniformity is about as unstructured as you can get. In and of itself, Clausius's description of the second law isn't terribly illuminating. You might, if you want to be *rude* about it, even go so far as to say that it's obvious.

Fortunately, 20 years after Clausius defined the Second Law, Ludwig Boltzmann decided to help us out by defining a concept known as entropy. No need for equations; a classic example will suffice.

Flip 100 identical coins. I'll wait. Provided that they are fair coins, that is, equally likely to fall on either side, you wouldn't be too surprised to find that about 50 of the coins (give or take) land on heads (H), and about fifty (take or give) on tails (T).

Provided sales of this book are robust enough (and I'll tell you *how* robust they need to be in a moment), one of you might look down at your coins and find, lo and behold, that all 100 of them are heads. You'd be so surprised! Wouldn't you?

A nerdy friend might make some snide comment about how you shouldn't really be *that* surprised to find all of your coins coming up heads. Wouldn't you, arguably, be just as surprised to find that the coins had come up with any particular pattern? I flipped 100 coins and got this:

HTTTHHTTTTHHTHHTTHTT
THTHTHHTHHHTHTHHHHTT
TTHHHTHTTHHHHHTHHHHT
TTTTTTHHHTTHTHTHHTHT
HHTTHHHTTTTHTHHTTTTH

Any chain of events is unlikely. You're 50 percent likely to get a heads on the first flip, and 50 percent likely to get a tails on the next flip, combining to 25 percent. To be clear, that is a 25 percent chance of flipping a *specific* pattern (HT for instance) but it's equally likely that you flip a similar combination like TH. Following this through to its natural consequence, any *particular* sequence of 100 flips is going to be improbable at the level of about 1 in 10^{30}, which is about how many of you need to be reading this book and flipping coins to have a decent shot of one string of 100 heads. Why get worked up by a *particular* unlikely sequence?

Did a friend or, god forbid, you, ever fall in love, get all dopey, and recount how incredibly unlikely it was that he or she found their soul mate *out of all the people in the world*? Or if you're even more self-centered, did you ever think about how unlikely it is that you exist at all? Besides the general question of life arising in the first place, there's the utter improbability of your parents meeting, and your two pairs of grandparents, and your four pairs of great-grandparents, and so on for tens of millions of generations. Seriously, what are the odds?

Of course any *particular* set of events was unlikely, but something had to happen. It's only with historical perspective that we start assigning meaning to things. Likewise with the coins, any *particular* pattern of heads and tails is incredibly unlikely. But a huge number of head–tail patterns have something in common: For 100 flips, there are approximately 50 heads and 50 tails.* The exact sequence of flips is known as the *microstate* of the system, while the overall measurement—in our case, it's the total number of heads, but it might just as easily be something like the temperature or density of a gas—is known as the *macrostate*. Getting all heads is unique. There's only one microstate for that particular macrostate, and so it's pretty special.

* My random sequence happened to be 50–50 exactly, but I wouldn't have been surprised to find it 48–52 or something like that.

Entropy is basically the number of microstates* in which the particles or coin toss results can be arranged to give you the same macrostate configuration. What, exactly, constitutes a unique macrostate is kind of complicated for systems that are more sophisticated than coin flips. Fortunately, (1) this isn't a mathematical textbook and (2) for most purposes, knowing exactly how to distinguish a macrostate doesn't really affect the gist of the argument.

Consider poker. There are approximately 2.6 million possible five-card combinations that you can draw out of a standard deck. Only 4 of them (1 for each suit) are royal flushes, the top hand in poker. However, there are over 1.5 million ways to draw a "high card" hand (no pairs, flushes, or straights). The type of hand (royal flush versus high card) is the macrostate, whereas the particular sequence of cards is the microstate. High card is a much higher entropy situation than royal flush. Or, conversely, a royal flush has much higher order. But you probably already knew that.

Instead of coin flips or cards, suppose you had four air molecules and you put them into the left side of a box. This is a very tidy, very low entropy way to arrange things. Allow nature to do its business, and the molecules will fly around, each spending about half its time on the left (L) and half the time on the right (R).

At any moment, you'll see a randomized snapshot of the four molecules. There are sixteen ways to arrange the molecules, but only two of them (LLLL, RRRR) have all four of the molecules on one side of the container. That's only a 12.5 percent probability. The rest of the time, the atoms are distributed more uniformly. For instance, there are

* Technically, the *logarithm* of the number of different permutations, but because we're not going to do any actual calculations, this detail shouldn't concern you too much. All that matters is that for macrostates that can be made in lots of different ways, the entropy is high. For macrostates that can be made with a only few microstates, the entropy is low.

ENTROPY

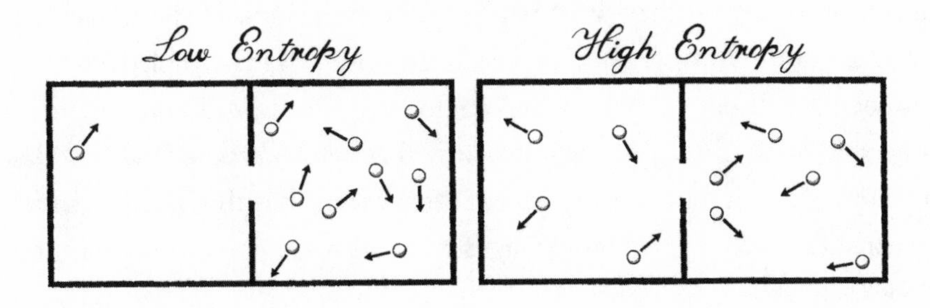

six ways (37.5 percent of the total) to sort the molecules so that there are *exactly* two on each side of the box. A uniform distribution is a higher state of entropy than is a concentrated one.

You could play this same game by taking the coins from before and tossing them into the air. For every head, it's like a molecule is in the left-hand side of the room, and vice versa. Do this a bunch of times, and you'll see that the molecules are almost always nearly equally distributed. With 100 molecules, randomizing 10 times a second, it would take roughly a trillion times the current age of the universe before you'd expect to find all of the molecules in one side of the box.

Increase the number of air molecules to, say, 10^{28} or more (the number that might fit into a small room), and probability dictates that random motions will ultimately make the molecules spread out evenly. Putting numbers on it, the two halves of the room will typically be equally filled to about one part in a hundred trillion.

At some point, systems become so large that it becomes not merely unlikely but nearly brain-bendingly impossible that entropy will decrease. That's why even though we call it the Second *Law* of Thermodynamics, it's really just a very, very good suggestion. So if you are a time traveler who loses his way, all you have to do to figure out the difference between future and past is to identify when the entropy is increasing.

This can't go on forever.

If the entire universe is just a big box of gas, it will eventually reach equilibrium, the point at which entropy is really at a maximum and the gas is split exactly fifty–fifty between the two sides. If the universe is at maximum entropy, all it can do is go down. As the molecules keep bouncing around, every now and again, a few extra will accumulate on one side or another, and the entropy will actually *decrease.* The fact that entropy exists in our universe doesn't just mean that things fall apart; it means that there are *lots* more ways for things to go wrong than for them to go right.

Consider a more conventional description of entropy, one dealing with temperature. In real gases, some molecules are flying around faster than others. The fast-moving ones are hotter than the slower ones. The highest entropy condition for our gas will require spreading out the temperature of the gas as evenly as possible. There are just more ways to spread the wealth than to keep it concentrated.

Those creationists among you may use this as evidence that complicated things—people or dinosaurs, perhaps—couldn't ever form in the first place. After all, you are a very highly ordered person, with an enormously complicated structure. Although your atoms could be rearranged quite a bit and still form a you-looking you, there are far, far more ways that your atoms could be rearranged to make something that looks nothing like you. If I take your exact chemical constituents and pour them into a concrete mixer, it's unlikely that we can pour your clone out the other end.

The good news is that we'll figure out how to turn those chemicals into your clone later in the book. The bad news, as it happens, is that it requires destroying the original (you) in the process.

Ultimately, it's not so surprising to find little patches of low entropy here and there, but there's a trade-off.

Entropy will increase over the whole universe. So, for example, if you make a nice refrigerator full of cold air, you'll do so at the expense of making a lot of high-entropy hot air. This is why your air-conditioner needs an exhaust but your space heater doesn't. It's also the reason you can't build a perpetual motion machine. As Carnot taught us, some of the energy will always be converted into heat.

CAN WE CHEAT THE SECOND LAW?

In the nineteenth century, James Clerk Maxwell devised a cool thought experiment to cheat the second law. Maxwell imagined a box filled with air molecules, some moving faster and some moving slower, but thoroughly mixed in a high-entropy state. In the middle of the box is a partition separating the left from the right, with a small trapdoor.

Whenever a cold molecule (one moving slower than average) approaches the trapdoor from the left side of the box, a very clever demon opens the door and lets the molecule through to the right side of the box. Likewise, whenever a hot molecule approaches from the right,

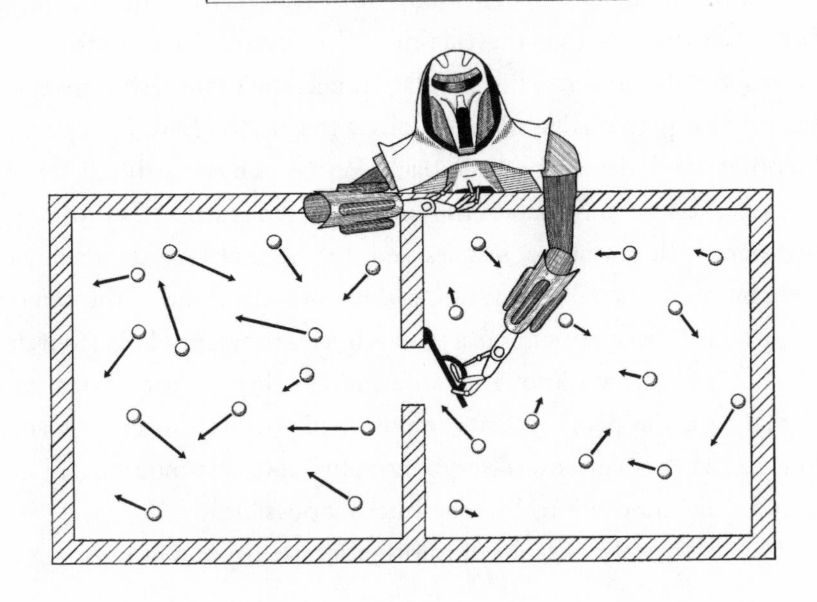

the demon opens the door and lets the molecule go into the left side of the box. Otherwise, the door remains closed.

Sounds simple, but if such a thing were possible, you'd literally never need to spend a cent on air-conditioning ever again. The demon does no work but still manages—McDLT-style—to make the left side hot and the right side cool.

I first saw this problem when I was an undergraduate and was profoundly unimpressed with it. Who cares about a few atoms here and there? Besides, if the Second Law is really only statistical in nature, does it really matter if we can circumvent it?

Yes, younger me. It does.

Remember how the Second Law is *supposed* to work: Heat should flow from hot regions to cooler ones. You probably didn't even need a science book to tell you that. Most of the energy used by our machinery gets wasted as heat, which is why we need to continuously burn coal, petroleum, or natural gas. If we could somehow employ a few million of Maxwell's Demons to recover the heat into useful energy, we'd have perpetual motion!

I have gotten more than my fair share of letters from amateur physicists with theories that they claim will revolutionize everything we know about the universe. It is standard procedure to throw out anything that predicts perpetual motion or violates the Second Law. Maxwell, on the other hand, deserves a little slack. Perhaps he really did discover a back door to somehow reduce the entropy in the universe. If you simply can't handle the suspense, rest assured, the Second Law stands unviolated, but to understand why, we need to enter the mind of the demon.

In 1948, Claude Shannon, a research scientist at Bell Labs, founded a branch of research known as information theory. Just as quantum mechanics made all of modern computing physically possible, information theory revolutionized cryptography and communication and helped make innovations like the Internet possible.

One of the major results of information theory is that information and entropy are intimately related. Just as the entropy of a gas describes the number of ways the atoms can be interchanged with one another, the information of a signal describes the number of different messages that can be transmitted.

Suppose I send a message that is exactly two characters long. I *could* in principle send you 26 × 26 = 676 different messages, but most of those letter combinations would be completely meaningless. Only a few (the Scrabble dictionary lists 101) are actual words.

To the computer scientists among you, this means that while in principle it would require about 10 bits (the 1s and 0s that are used to store data) to differentiate between every *possible* two-letter combination, if you know that you're transmitting a word, you only need about 7 bits. What a savings!

Communications can be significantly compressed by noticing that certain letters are used less frequently than others. *E*s, for example, show up way more often than *Z*s in the English language. If you're playing hangman, simply knowing that a *Z* appears in a word dramatically decreases the number of possibilities. This is why the former is worth only one point in Scrabble and the latter is worth ten, and why the letter *E* in Morse code is

• •

whereas the letter *Z* is

— — • •

Z takes far longer to tap out, but that's okay, because you're going to do it far less frequently. The more complicated (or unlikely) a message is, the more information it carries, and the more bytes of data you'd need to store it on a computer.

Which brings us back to Maxwell's demon. Let's take neuroscience out of the equation and suppose the demon is really some fantastical metal man who stores his data digitally. The memory in a computer is really just a series of 0s and 1s. Whether or not we have any files on the disk, there are a finite number of different patterns that *could* be stored. That number can be computed by taking 2 × 2 × 2 . . . multiplying another two for every bit on the disk. More bits means more possible different patterns.

Every time the demon has to decide whether or not to let an atom through his trapdoor, he takes a measurement and makes a recording of the speed of the atom. Suppose, further, that he has a special (but very tiny) disk set aside entirely for the purpose of storing the speed of an atom long enough to decide whether it should be permitted through the partition. At the beginning of the experiment, all of the registers in the disk are set to zero, a very low entropy configuration.

On the other hand, if a disk is filled with a seemingly random array of 1s and 0s, there's either a hell of a lot of information stored on the disk, or the disk is entirely random and there's a huge amount of entropy. But for our demon, we're going to start with a clean slate.

The first atom approaches the door, and the demon makes a measurement, dutifully stores it to disk, and decides to let it through. A second atom approaches and—what's this?—the disk is completely full from the first measurement. The demon has no choice but to delete his original entry before continuing on, and there's where Maxwell's thought experiment ultimately breaks down.

In 1961, the computer scientist Rolf Landauer showed something remarkable: If you delete a bit of information, you necessarily create an equivalent amount of entropy in the universe. Creating—and ultimately deleting—a record of the motions of the atoms releases at least as much entropy as the demon was supposed to have saved by partitioning the atoms in the first place. The demon doesn't really decrease the overall entropy of the universe by playing his gas games. He just redistributes it.

It's worth spending a few moments more on the memory of the demon before we move on. In making measurements of atoms or of anything else, we made the basic assumption that the demon started with a blank slate, perhaps initially configured to all zeros. But what if we *didn't* make that initial assumption?

There are clearly some configurations of memory that are more special than others. Much like with the Scrabble letters, we could simply recognize that most combinations of 1s and 0s are garbage, but even so, picking tiles out of a bag will occasionally give us a real word. The problem is that a randomly generated (but otherwise meaningful) sequence on the board looks just as real as a word that somebody played intentionally.

If you found a disk drive with lots of seemingly random 1s and 0s, you'd quite reasonably assume that all of those bits represented real data stored on the drive. By the same token, if you were a robot who read the disk and found a complicated series of numbers, you'd assume it was real data. From a robot's point of view, data on the disk are *exactly* the same thing as memory, and there's no distinguishing between a real memory formed from experience and one formed by the equivalent of throwing down random tiles.

In other words, we generally assume that whatever complicated patterns are in our brain or on a Scrabble board or in the physics of the universe are somehow an accurate reflection of what really happened in the past.

The philosopher David Albert described the Past Hypothesis as the *assumption* that the past was at lower entropy than the present.* For a computer, this might mean that we start by zeroing out the registers and adding data as we go along. If the Past Hypothesis is correct, then

* And this is an assumption. It is entirely possible (though, I think, unreasonable) to suppose that the dinosaur bones in your backyard were put there recently rather than 65 million years ago through the death of an actual dinosaur.

information encoded into memory is a realistic representation of what happened in the past. If the computer disk really started at high entropy, we'd have no idea which were real memories and which were noise. For us to make any sense of the past, we have to assume that we, and the universe more generally, started at extremely low entropy.

And *that* brings us to an extremely vexing question about the early universe. . . .

WHY DID THE UNIVERSE START OUT SO BORING?

You sit in a hot bath, and at first, you feel all warm and cozy, but then things take an unfortunate turn: The water in the room starts to equilibrate with the air and you get cold and shriveled.

The same holds for the future of the universe. As time goes on, heat gets more and more evenly distributed in the universe. Stars burn out, black holes ultimately evaporate, and everything goes dark and cold. The ultimate condition of the universe will be as a uniform, incredibly vast and incredibly cold pool of photons.

And what of our origins? Back in Chapter 1, we saw a microwave map of the early universe, with little splotches of hot and cold. What I didn't mention at the time was that the "hot" spots are only about 1 part in 100,000 warmer than average, and the "cold" spots only about 1 in 100,000 cooler.

The end of the universe and the beginning sound remarkably similar, yet I'm claiming that the end of the universe will be in high entropy, whereas the beginning was low entropy. What gives?

The key is gravity. Start with a perfectly smooth universe, and add only a few clumps where the density is just a little bit higher than average. Before you know it, some of the nearby matter will fall in until the little clump becomes a bigger clump.

Entropy isn't just a measure of how many ways you can mix up a system and have it look the same. As we saw in radioactive decay, everything wants to evolve to the lowest energy state possible.* As the molecules fall into the clumps, the energy gets turned into heat, and the heat is essentially entropy. Entropy increases as tiny clumps become much larger clumps, ultimately resulting in galaxies, stars, and you.

Gravity was far more important in the early universe when everything was packed closer together than it is today. Local gravity is far more important now than it will be in the distant future. To a universe dominated by gravity (like the beginning of time), a perfectly smooth distribution is the *lowest* entropy configuration. In the future, when gravity becomes less important, a perfectly smooth distribution is the *highest* entropy configuration.

We can see the effect of gravity in the distribution of galaxies. Beginning in 2000, the Sloan Digital Sky Survey (SDSS) started mapping out much of the nearby universe. Distances were measured for over a million galaxies, and images were taken of over a hundred million more. There is clear structure, in the form of clumps, filaments, and empty regions known as voids. However, when we look back in time (or equivalently, far away), the universe appears incredibly smooth.

This is a serious puzzle, and one that, in many ways, is related to the question of why the arrow of time flows one way and not the other. Take the universe as it is now, and imagine a movie leading up to the current state of affairs. If we run that movie in reverse, everything will start off at high entropy and end up in a state of low entropy. It must; as we've seen, the laws of physics are essentially time reversible.

Take it a step further and change the present universe ever so slightly. Move a few atoms around here and there. If you run this (slightly altered) universe backward, you won't end up at the "beginning" with a smooth universe at all. The odds of finding the beginning

* And you have to put the energy *somewhere*. The First Law of Thermodynamics is that energy is conserved. Also, that you *do not* talk about thermodynamics.

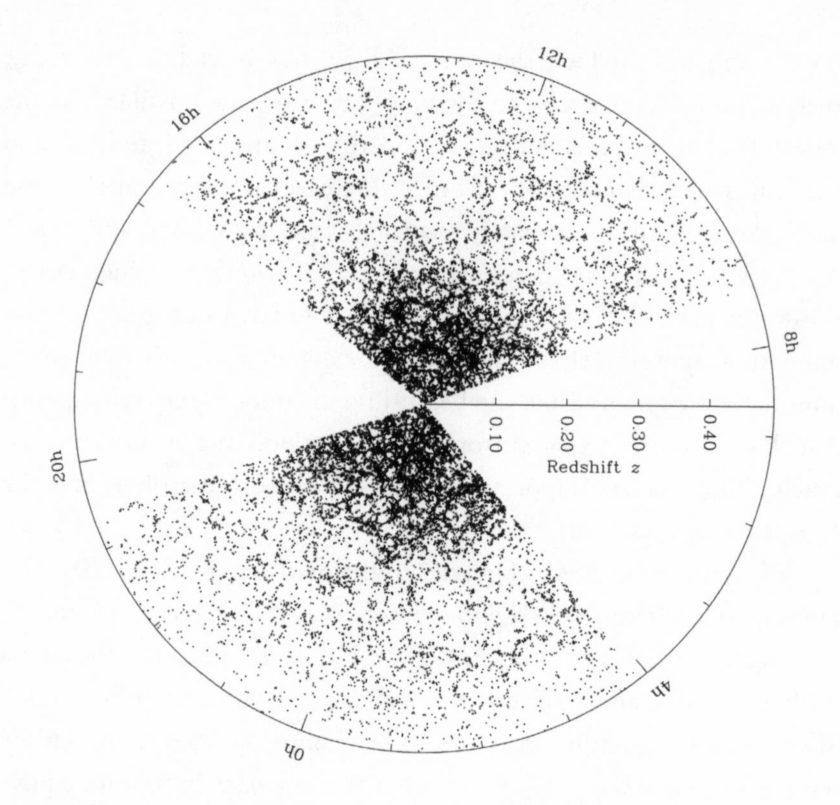

of the universe in a low-entropy condition are ridiculously improbable—as improbable as the possibility that the universe will evolve to a low-entropy state.

It's a bit difficult to even define what I mean by *improbable* in this context. Normally, when we say something is unlikely, we mean there is a series of events that lead up to a final event, and we base the probabilities on previous events. For the beginning of the universe, there *is* no previous condition.

This is the Past Hypothesis writ large. We might even imagine that it's a law of nature; perhaps low-entropy early universes are just the rule. Honestly, though, that's not very satisfying. This is still an open question, but a few ideas have been floated that are more interesting than "The Universe started off at low entropy because it did."

For example, it may be that ours isn't the first universe. A number of people, including Princeton University physicist Paul Steinhardt, Neil Turok, and their colleagues, have suggested that the universe undergoes periodic phases of expansion. One of the properties of the Ekpyrotic Universe,* as it's known, is that any given patch of the universe gets stretched and stretched over time. Entropy doesn't decrease in this universe, but as the universe stretches, the entropy within a small patch *can* become more dilute. Our universe may simply be a small patch of a much, much larger "multiverse," one in which the overall entropy is and remains enormous.

Others have a different spin on the role of the multiverse. Caltech physicist Sean Carroll describes time as an emergent. He argues that the flow of time in our universe—and every other bubble—is the increase in entropy:

> The arrow of time isn't a consequence of the fact that "entropy increases to the future"; it's a consequence of the fact that "entropy is very different in one direction of time than the other."

Others have gone even further. Dutch physicist Erik Verlinde, for example, claims that even seemingly fundamental phenomena like gravity come out of the Second Law of Thermodynamics and string theory.

While these ideas are intriguing, they are not necessarily a consensus view of physics. For my part, I am somewhat skeptical. We will talk a fair amount about the multiverse in the next chapter, but part of the difficulty in talking about it is that it's not clear that we'll ever have any direct observational or experimental evidence that there are bubble universes.

Given the choices available, my personal preference is to suppose

* From the Greek, literally, and awesomely, "conversion into fire."

that the beginning of the universe was low entropy just because that's how it was. I mentioned earlier that when you're talking about the beginning of time, measures of probability start to lose all meaning, so when people talk about how staggeringly unlikely our low-entropy beginnings are, it's not clear what we should expect. Richard Feynman put it pretty well:

> [A low-entropy past] is a reasonable assumption to make, since it enables us to explain the facts of experience, and one should not expect to be able to deduce it from anything more fundamental.

This is the problem with talking about first causes. We can't deduce what the law is because, as near as we can tell, there really was only one beginning of time. Even though T Symmetry demands that the laws of the universe are microscopically reversible, at the end of the day, there must be a single and unique direction that leads us back to the beginning of the day. The not-so-obvious symmetry of time points us uniquely back to the beginning.

Chapter 3

THE COSMOLOGICAL PRINCIPLE

IN WHICH WE LEARN WHY IT IS DARK AT NIGHT

There are two things I hope I've made painfully clear. The first is that stoner questions are usually far more perceptive than they might appear at first glance. The second is that it is incredibly important to remember how utterly mediocre we are. Those two realizations can take us dangerously close to solipsism. How do you know you're not a brain in a jar, with the entire universe simulated entirely for your benefit?

This isn't (just) meant to be facetious. It is easy—too easy—to imagine that you're at the center of the universe, both metaphorically and literally. Go outside and watch the sky. The sun, the planets, and the stars appear to rotate around us. Your expectations of the universe depend a great deal on your sense of self-importance.

Your assumptions are so ingrained that you may not even recognize another truly profound question when you hear one. If I ask, "Why is the sky dark at night?" an answer from the man on the street is likely to be impatient and something along the lines of "Because the sun is on the other side of the earth, dummy." Well, that is both mean and wrong. The darkness at night is far less obvious than you might first suppose. Both the depth of the question and its resolution are explained by symmetries.

THE CENTER OF THE UNIVERSE

To the ancients, the dark night sky posed no real problem. This stems, ultimately, from a complete misunderstanding of how the universe actually works. As I mentioned, Aristotle in particular was wrong about nearly everything he wrote about the physical world: the nature of the five elements,* the working of gravity, and worst of all, his idea that the sun orbits the earth. Let's just say he was much more on the mark when it came to ethics.

His mistakes are, to some degree, understandable. Physics as a field of study was so new in the fourth century BCE that Aristotle coined the word *phusikes*, at least in the sense that we use it. And almost everything that Aristotle wrote about the physical world *feels* as if it should be true.

Heavy objects, for instance, typically *do* fall faster than light ones, but only because air resistance is relatively less important. The sun and stars certainly *look* as if they are in orbit about the earth.

The sun and the earth are in mutual orbit around a point about 450 kilometers from the center of the sun. The sun actually sort of wobbles around that point, a detail that tends to get overlooked. The sun, the earth, and the rest of the solar system are in orbit around the center of the Milky Way Galaxy, roughly 27,000 light-years away, and the whole galaxy is flying around the Virgo Supercluster, a region over 100 *million* light-years in diameter.

---

* Earth, water, fire, air, and aether. This last was the substance that made up the stars and the celestial spheres.

That said, Aristotle was intuitively right about one thing: The universe really is governed by underlying symmetries. From his *Physics*:

> In circular motion there are no such definite points: for why should any one point on the line be a limit rather than any other? Any one point as much as any other is alike starting-point, middle-point, and finishing-point, so that we can say of certain things both that they are always and that they never are at a starting-point and at a finishing-point.

There's something special about a circle and, by extension, a sphere. No matter how you turn it, it looks the same. And *everything* was spherical with this guy. The stationary earth was a perfect sphere. It sat in the middle of a series of about fifty or so perfectly concentric spherical shells containing the moon, the planets, the sun, the distant stars, and ultimately, the unmoved mover himself.

But wait! Didn't the ancients think that the world was flat? Nope. The ancients were wrong about lots of things, but surprisingly, they got this one right.

Going back at least to Pythagoras (the triangle guy) in the sixth century BCE, the consensus view among people who paid attention to this sort of thing was that the earth was round. As Stephen Jay Gould put it:

> There never was a period of "flat earth darkness" among scholars (regardless of how the public at large may have conceptualized our planet both then and now). Greek knowledge of sphericity never faded, and all major medieval scholars accepted the earth's roundness as an established fact of cosmology.

Round, but not perfectly so. In the late seventeenth century, Isaac Newton showed that the earth is an oblate spheroid, slightly larger around the equator than through the poles. He made this argument on purely theoretical grounds and it wasn't until early in the next century, when several expeditions were sent out to actually *measure* the earth, that it became clear that the earth was not spherical. The earth is about 30 kilometers higher around the equator than through the poles.

Perfect spheres or no, Aristotle's model was deeply flawed from simple observations. It had been known for quite some time that planets didn't simply move through the night sky in one direction; occasionally, they crossed back on themselves. What's more, they seemed to vary in brightness throughout their cycles in a fairly complicated way. Neither of these effects could be described by Aristotle's spheres.

In the second century, Claudius Ptolemy perfected the Aristotelian geocentric model in his *Almagest* (literally, "The Greatest"*). Ptolemy introduced a universe in which the planets moved on epicycles—a sort of circle upon circle motion—that gave surprisingly accurate predictions of where planets will appear in the night sky. The Ptolemaic model quickly became both scientific and religious orthodoxy and continued to be so until the seventeenth century.

You might excuse the ignorance of the ancients by supposing that it had simply never occurred to anyone that the sun was at the center of the universe. But more than 400 years before Ptolemy, the Greek astronomer Aristarchus of Samos† described a heliocentric universe. Very little of Aristarchus's work has reached us directly, but it's been described by others, including the mathematician Archimedes:

---

* I *know*. Seriously.

† We saw Aristarchus before. He gave us one of the best ancient estimates of the distance to the sun. This is no coincidence. The sun appears relatively large in the sky. If it is far away (and it is), then it must be enormous—far larger than the earth. If the sun really is bigger than the earth, then it stands to reason that the earth orbits the sun rather than the other way around.

> Aristarchus has brought out a book consisting of certain hypotheses . . . that the fixed stars and the Sun remain unmoved, that the Earth revolves about the Sun on the circumference of a circle, the Sun lying in the middle of the Floor, and that the sphere of the fixed stars, situated about the same center as the Sun, is so great that the circle in which he supposes the Earth to revolve bears such a proportion to the distance of the fixed stars as the center of the sphere bears to its surface.

Aristarchus's model was simple and fairly accurate, certainly consistent with the measurements of the day. It had an enormous shortcoming, however; it implied that humans were somehow ordinary—that stars like the sun were everywhere and that the earth wasn't at the center of the cosmos.

You may already know the rest of the story. The Ptolemaic model was adopted by the Catholic Church, and it became heresy to claim that the earth was anywhere other than the center of the universe.

It wasn't until 1543, when the Polish astronomer Nicolaus Copernicus published his *De revolutionibus orbium coelestium*, that Aristarchus's model was rediscovered. Copernicus kept out of trouble* by writing a rather weaselly introduction† to his book:

> Those who know that the consensus of many centuries has sanctioned the conception that the earth remains at rest in the middle of the heaven as its center would, I reflected, regard it as an insane pronouncement if I made the opposite assertion that the earth moves. . . . I thought that I too would be readily permitted to ascertain whether explanations sounder than those of my predeces-

* Burning at the stake, torture, arrest—that sort of thing.

† There is some scholarship suggesting that the introduction was actually written by his assistant, in which case, I apologize for likening Copernicus to a weasel.

> sors could be found for the revolution of the celestial spheres on the assumption of some motion of the earth.

In other words, "Don't worry. This is all just a mathematical exercise and doesn't necessarily tell us anything about the real universe." He also had the good sense to write in Latin (to keep it out of the hands of common folks) and even more sensibly to die shortly after publication.

Not everyone was so well advised. Giordano Bruno, who was first and foremost a Dominican friar, went much further than Copernicus. Not only did Bruno argue that the sun was the center of our universe, but he also thought (correctly, as it turns out) that all of the stars are simply suns like our own. He wasn't framing things in hypotheticals. Instead, he argued:

> In space there are countless constellations, suns and planets; we see only the suns because they give light; the planets remain invisible, for they are small and dark. There are also numberless earths circling around their suns, no worse and no less than this globe of ours. For no reasonable mind can assume that heavenly bodies that may be far more magnificent than ours would not bear upon them creatures similar or even superior to those upon our human earth.

Bruno was right about the ridiculous number of planets in the universe. As of this writing, there are over 800 known planets or planetary candidates in our galaxy, and if the early results from the Kepler planet-finding satellite mission are any indication, it looks as though there may be a great many that are potentially habitable. Being right, it turns out, isn't always enough. In 1600, the Inquisition burned Bruno at the stake for heresy.

Eventually, the evidence for a heliocentric universe became incontrovertible. In 1609, Johannes Kepler published his *Astronomia nova*, which, among much else, laid down the laws of planetary motion.

Kepler was a student of Tycho Brahe, one of the greatest observational astronomers of the day. Tycho* (as he's almost always called) had an entire island dedicated to his observations and had made some of the best measurements of the motions of the planets. His goal was to prove a sort of hybrid Ptolemaic–Copernican model wherein the sun orbits the earth, but the rest of the planets orbit the sun.

Kepler went into Tycho's service essentially just to get his hands on the data. Upon Tycho's death in 1601, Kepler went to work. As he later described it:

> I confess that when Tycho died, I quickly took advantage of the absence, or lack of circumspection, of the heirs, by taking the observations under my care, or perhaps usurping them.

It was a good thing, if a little unseemly. From these observations, Kepler deduced that the orbits of the planets are ellipses, not circles.

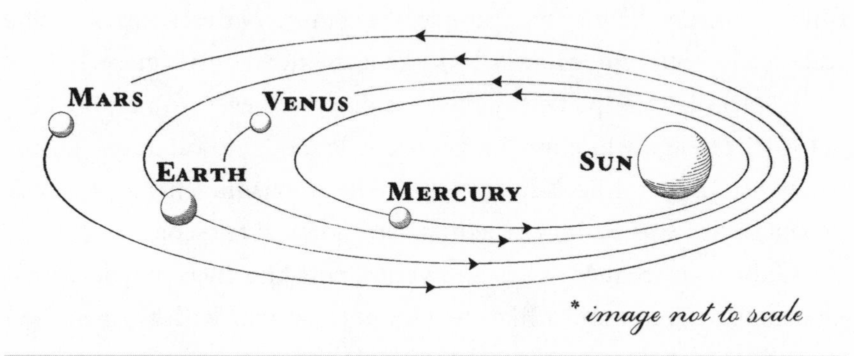

* He also lost his nose in a duel with the fancifully named nobleman Manderup Parsberg. The (almost certainly apocryphal) story was that they argued about who was better in math and decided to settle it with swords. Tycho had his nose replaced with a prosthetic made of silver and gold. That's not really relevant to our story so much as it gives you an idea of the badassery of old-school astronomers.

In most cases, the ellipses are *very nearly* circles. Although the earth is, on average, about 93 million miles from the sun, it's about 5 million kilometers closer to the sun at perihelion (in early January), than during aphelion six months later. The particularly astute among you may notice that in the Northern Hemisphere, we're actually *closer* to the sun in the winter than the summer. Suffice it to say, the distance to the sun is *not* the origin of the seasons.*

Kepler took a long while to stumble into elliptical orbits. For one thing, he assumed that if planets followed such a simple orbit, then some earlier astronomer must surely have discovered it. For another, he hit a number of dead ends. One of the most interesting came in his awesomely titled *Mysterium cosmographicum* in which he proposed that the orbits of all of the planets were determined by inscribing each of the Platonic solids in one another like an astronomical turducken.

The orbits of the six known planets (everything out to Saturn) could be approximately fit by putting a cube inside a tetrahedron inside a dodecahedron inside an icosahedron inside an octahedron inside a sphere. This is yet another example of scientific inspiration coming from symmetry. It's also an example of a symmetry that turns out to be completely irrelevant. He was lucky that the fit was even close.

Kepler's laws of planetary motions described everything perfectly, but it still took a while for the heliocentric model to catch on. Fortunately, in 1609, while Kepler was finally publishing his *Astronomia nova*, Galileo was building the first astronomical telescope.

Galileo saw that Mercury and Venus, now known as the innermost planets, have phases, just like the moon, consistent with their motion around the sun. He saw that the Milky Way was

* It's all about the tilt of the earth. The Northern Hemisphere spends more time facing the sun in June, July and August, and in the Southern Hemisphere, it's reversed. You'd be surprised how often people get this wrong.

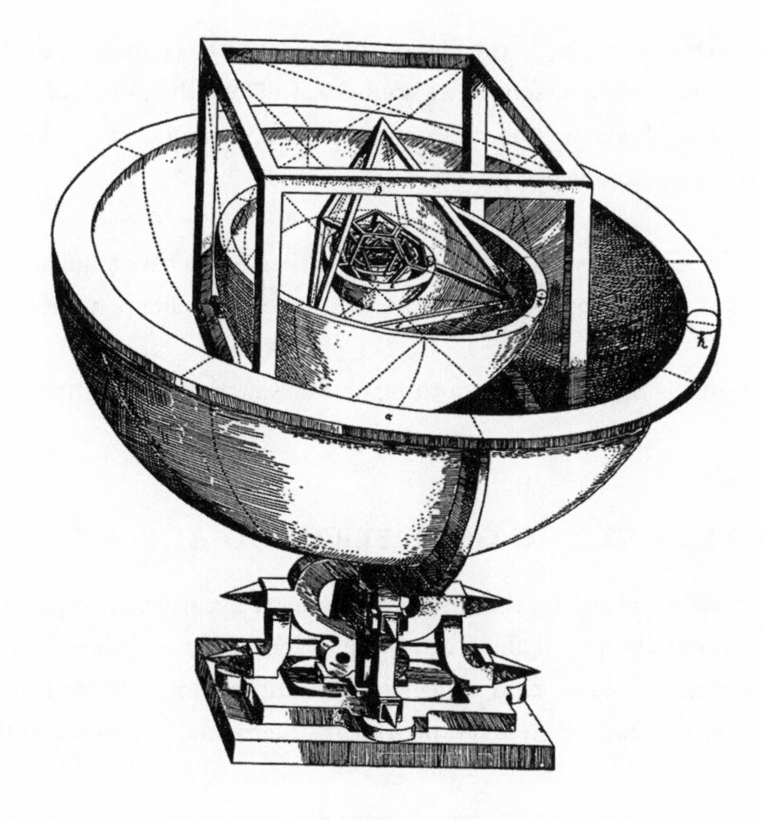

> nothing else but a mass of innumerable stars planted together in clusters.

But most tellingly, he saw that Jupiter appeared to have a number of satellites:

> There are three stars in the heavens moving about Jupiter, as Venus and Mercury around the sun.

If Jupiter could be the center of its own little system, how could earth be the center of everything?

Things went somewhat better for Galileo than for Bruno. Galileo was merely forced to recant and sentenced to house arrest for the remainder of his life. By the end of the seventeenth century, the belief in other planets was essentially a nonissue. In 1698, the Dutch physicist Christiaan Huygens who, among much else, was one of the first to describe light as a wave, put it:

> Why [should] not every one of these stars and suns have as great a retinue as our sun, of planets with their moons to wait upon them?

And nothing bad, or at least nothing church sanctioned, ever happened to him.

WHEREVER YOU GO, THERE YOU ARE

Copernicus was among the first to realize a greater truth: There's nothing particularly special about our place in the universe. This is a lesson that humanity was forced to learn again and again. Our mediocrity extends much farther than our own solar system. Galileo observed that

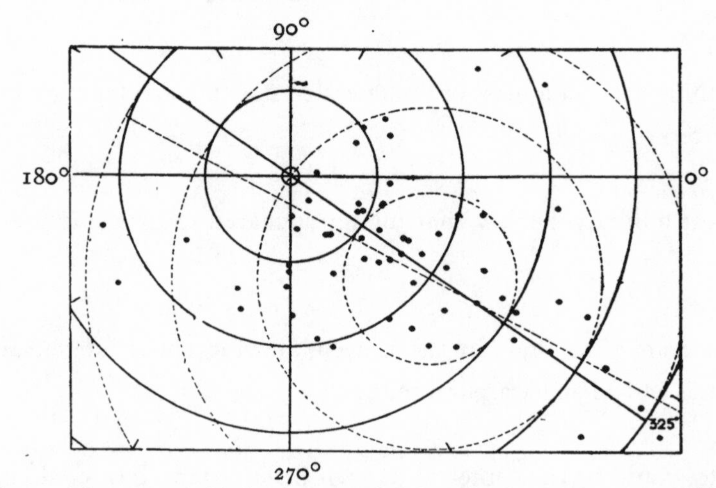

FIG. 5. The system of globular clusters projected on the plane of the galaxy. The galactic longitude is indicated for every thirty degrees. The "local system" is completely within the smallest full-line circle, which has a radius of a thousand parsecs. The larger full-line circles, which are also heliocentric, have radii increasing by intervals of 10,000 parsecs. The broken line indicates the suggested major axis of the system, and the broken circles are concentric about its center. The dots are about four times the actual diameters of the clusters on this scale. Nine clusters more distant from the plane than 15,000 parsecs are not included in the diagram.

there are countless stars out there, each with an equal claim to being in the center of the universe.

In 1918, the astronomer Harlow Shapley made a map of sixty-nine globular clusters in the Milky Way. These are very tight bundles of a hundred thousand stars or more, and it was reasonable to suppose that the globulars were distributed symmetrically around the center of the galaxy. He found that even within our own Galaxy, we're not in an especially privileged place. We're just one of about 10 billion star systems out in the provinces.

As Douglas Adams put it:

> Far out in the uncharted backwaters of the unfashionable end of the Western Spiral arm of the Galaxy lies a small, unregarded yellow sun.
>
> Orbiting this at a distance of roughly ninety-eight* million miles is an utterly insignificant little blue-green planet whose ape-descended life forms are so amazingly primitive that they still think digital watches are a pretty neat idea.

That's not nearly the end of it. In the 1920s, Edwin Hubble showed that our galaxy is just one of an enormous number of island universes flying through space. As we've seen, the SDSS has mapped over a hundred million galaxies, but a conservative estimate would put the total number in the observable universe in the low trillions. Taken on average, those trillions of galaxies seem to be distributed remarkably uniformly throughout space. Put in the language of symmetry, the universe is *homogeneous*. Likewise, the universe seems to be more or less the same in the Northern Hemisphere as in the Southern. Again, to break out the technical terms, the universe appears to be *isotropic*.

---

* Adams wasn't an astronomer and was, at any rate, a Brit, so he can be forgiven for getting his unit conversion wrong. The number is closer to 93 million miles.

These observations form the basis for the so-called Cosmological Principle, which says, in essence, that the universe is more or less the same in all directions and in all places. While it's grounded in observations, the Cosmological Principle is really axiomatic. Just as the assumption that constant physical laws allow us to interpret the past and predict the future, the Cosmological Principle allows us to reasonably interpret data from other parts of the universe.

Our first real understanding of the universe outside of our Galaxy came courtesy of Edwin Hubble. As we've seen, he didn't just give us a scale of the universe; he also discovered that nearly every galaxy in the universe appears to be receding from us.

The expanding universe may leave you with the erroneous idea that the universe has a center. It doesn't. To understand why not, we need to talk a little bit about relativity. We've already seen that *Special* Relativity involves a close relationship between space and time. What makes *General* Relativity so awesome is the realization that gravity can *curve* either space, time, or both.

If you don't have an intuitive feel for curved space, don't feel bad. It's easy to get lost in the equations. Fortunately, the International Order of Cosmology Popularizers has come up with the perfect analogy, and if you promise not to take it too literally, I'll follow suit.

> Glue a bunch of little plastic galaxies onto a giant rubber sheet.
> Find a bunch of stout men to grab the sheet on all sides.
> Pull vigorously.

An ant living in one of the galaxies will consider herself the center of the universe, while all of the other galaxies in the universe appear to recede into the distance. What's more, the greater the distance between two galaxies, the faster they appear to be receding from one another, exactly the effect observed by Hubble.

I could drop you off in any galaxy, and provided you were

Expanding Universe as a Rubber Sheet

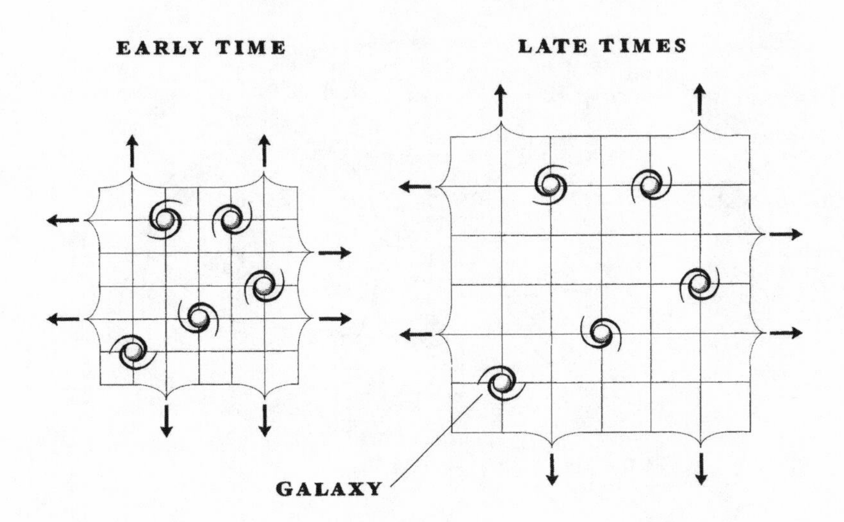

egocentric enough, you would assume yourself to be at the center of the universe. But, and this is the key point, *any* observer will see the same basic thing.

Run the clock of the universe back in time, and the distances between all of the galaxies will be on top of one another. *Where did the Big Bang happen?* It happened *everywhere.*

There is a danger in taking this analogy too literally. A particularly tenacious ant could make an adorable little spaceship and sail off to the edge of the rubber sheet, for instance.

In our (non-rubber-sheet) universe, on the other hand, there is absolutely no prospect of ever reaching the edge. Just as the universe has no center, it also has no edge. This leaves us with essentially two options. The first is, quite frankly, terrifying. It could be that the universe goes on literally forever. In this scenario, the universe is not just very, very big; it's infinite. Infinite!

We'll explore the practical difference between a gigundous universe

Toroidal Universe

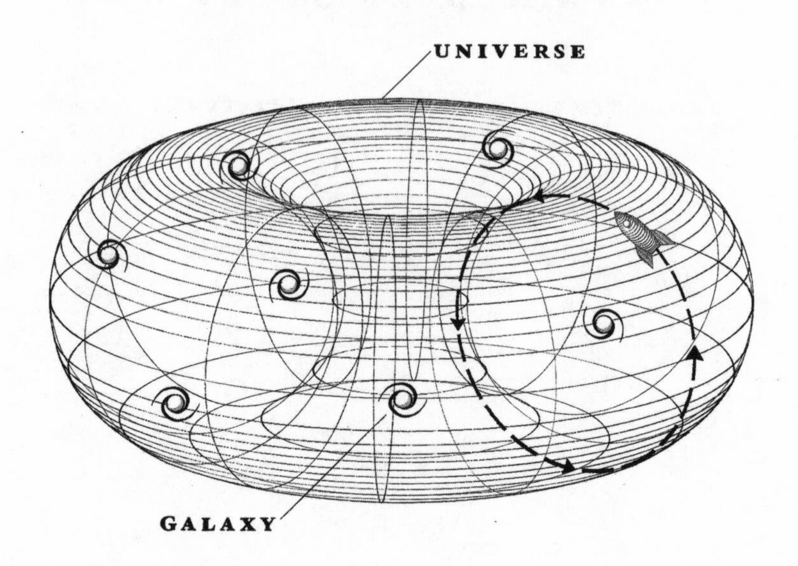

and an infinite one in a little bit, but for my own part, I'm much more comfortable with option two: The universe could fold back on itself.

It's equivalent to Pac-Man disappearing from one side of a screen and reappearing on the opposite side. From Pac-Man's perspective, he can just keep traveling and traveling and never reach the end.

Don't fret; the earth behaves the same way. Ignoring arbitrary man-made demarcations like the International Date Line, you could continue traveling east for all eternity, never hitting an edge or a center. You'd just pass the same points again and again.

From a practical perspective, there's not much of a difference between a repeating universe and an infinite one. The combination of the expanding universe and the speed limit of light conspire to prevent us from ever looping around the universe and ending up where we started. But that doesn't stop us from asking the question, How big is the universe, anyway?

UNIVERSE OR MULTIVERSE?

Space is big. Very big.

As to how big, we honestly can't say, ever. We can't see the entire universe because it's been around for only about 14 billion years and light travels only so fast. On earth, we call the maximum distance that we can see the *horizon*, and it's no different for the universe as a whole.

In principle, we could map out trillions of galaxies within this horizon, but that isn't necessarily all there is. There's the very real possibility that the universe beyond what we can see is different from what's nearby. Not only can't we see what's happening hundreds of billions of light-years away, because *everything* travels at the speed of light or slower, nothing outside the horizon could possibly have affected what happens here on earth.

Stranger still, because our universe is accelerating, it turns out that we won't simply see more and more distant galaxies as time goes on. Galaxies within our ultimate horizon are *only* about 60 billion light-years away. Whatever is happening beyond that distance will forever remain a mystery.

In almost every way that matters, everything outside of our horizon is a separate universe and this means that, like it or not, we're living in a *multiverse*, at least at some level. If you're schooled in science fiction,* the idea of a multiverse won't be that unfamiliar, but *multiverse* means different things to different people. Fortunately for us, MIT physicist Max Tegmark came up with a whole hierarchy of them. Just to be clear, beyond Level 1, which we're pretty sure about, this is a very speculative list, and becomes more and more speculative as we proceed. Keep in mind that for the moment, we're just organizing our thoughts.

---

* And of *course* you are.

Level 1 Multiverse: The Universe Is Big, But in a Normal Sort of Way

Every 100 billion light-year patch of space can, for all intents and purposes, be considered an island unto itself. But if each island is disconnected, there's a deeper question of why, and whether any given patch is likely to look like any other.

Believe it or not, this is actually a question that we can probe. But first an observational fact: We are surrounded by radiation from the beginning of the universe, and that radiation is uniform to about 1 part in 100,000. This fact is made all the stranger when you realize that light hitting us from above the North Pole and light coming from below the South Pole were emanated from very, very distant points in the universe. These two patches of light came from regions that don't seem to have *ever* been in thermal contact with one another.

This is one of the deepest and most troubling questions in cosmology. The early universe was very small, but it was small for only a short time. It doesn't seem like regions separated by more than about a degree on the sky should ever have had a chance to mix with one another, and yet, the entire sky looks remarkably uniform. This was, remember, one of the assumptions of the Cosmological Principle.

In the 1980s, Alan Guth, then at the SLAC National Accelerator Laboratory, proposed an idea called inflation to get around the horizon problem. And though it's going to sound a bit strange, I should tell you in advance that inflation is very much orthodoxy to most cosmologists. It explains an awful lot about the universe as we now observe it.

The first moments of the multiverse were pretty active, and the first 10^{-35} second, especially so. For a brief instant, the universe underwent a tremendous exponential expansion, growing small patches of space by a factor of 10^{60} or more.

If inflation is correct (and again, we're reasonably certain that it is), then there is a lot space beyond the space that we can see. Each bubble is its own universe, and you could easily imagine that if there were enough of them, a few of them might look a little like our own, perhaps exactly like our own. In most models of inflation, these bubbles spawn other bubbles and on and on forever, producing the infinite universe that freaked us out so much in the first place.

How big does a Level 1 multiverse have to be before we start getting exact duplicates of everybody? Pretty damn big. Tegmark estimates an identical universe at around $10^{10^{29}}$ meters from here, the biggest number that's going to come up in this book outside of infinity itself. What this means is that every atom in the duplicate universe is in precisely the same spot and moving with the same speed—up to the limits of quantum uncertainty—as in our own universe. That means that even if alternate you didn't have exactly the same history as your own, your (alternate) brain is configured so that you *think* you did.

See? We're back to evil twins!

If the universe were infinite, then it would be plenty big enough to accommodate not only a duplicate of you, but an infinite number of yous.

It's humbling, and also a bit creepy. It's like you have an infinite number of stalkers.

If the universe *weren't* infinite, then you could probably rest easy in your uniqueness. A conservative theoretical estimate puts the minimum size of our multiverse at around 10^{80} meters, which seems huge until you realize that it's only a tiny, tiny fraction of the space required for duplicates.

Level 2 Multiverse: Different Universes with Different Physics

Our bit of the universe grew out of one tiny patch of the very early multiverse, but as we've seen, ours is not the only bubble. What's more, some, perhaps all, of those bubbles may have physics just a wee bit different from our own. Electricity might be a bit stronger or weaker; the strong force (the one that holds protons and neutrons together) could be a bit different; there could be more than three dimensions.

Let me make a few things clear about Level 2 multiverses:

1. It's not obvious that this model is correct. It may be that the fundamental forces really are hardwired in nature and that all universes have the same underlying physics.
2. If there really is a Level 2 multiverse, these universes aren't going to look much like our own. A lot of them are going to be devoid of stars or galaxies, very nearly empty, or perhaps collapsed entirely under their own gravitational pull. Physics needs to be *very* finely tuned to make things like stars or heavy elements—or us, of course—and most universes simply don't pass muster.
3. The universe *still* doesn't have an edge. The transition from universe to universe isn't a brick wall. The universes inside a Level 2 multiverse are themselves potentially Level 1 multiverses.

And even Level 2 isn't the end of the story. Tegmark also has third- and fourth-level multiverses, which are arguably even *more* speculative and not that related to the symmetry question of whether the laws of physics are the same everywhere. But we're still going to talk about them because they're very cool.

Level 3 Multiverse: The Many Worlds of Quantum Mechanics

I've already said a little bit about how quantum mechanics works, and most physicists simply accept that there needs to be a little bit of randomness (or quite a lot of randomness) and the potential for weirdly entangled coherence between distant events.

Not everyone is so sanguine about this. In 1957, Hugh Everett, then working as a scientist at the Pentagon, came up with the Many-

Worlds Interpretation (MWI) of quantum mechanics. Everett wasn't creating a whole new set of physical laws. What he was saying in essence was "See all of these experiments that describe quantum behavior? Well, here's a different way of looking at them."

In the MWI, every time a quantum event can be measured in a random state, a new set of universes is created. In one universe, an electron is measured as spin-up. In another, it's spin-down. The curious thing is that, according to MWI, these universes can interact with one another, causing the strange behaviors of quantum interference.

As I said, mathematically, the MWI produces exactly the same expectations for quantum experiments as does the standard, Copenhagen, interpretation that most physicists (including me) subscribe to. But it also gives us a whole new way to think about the multiverse and one that, quite frankly, is outstanding if you're in the business of producing sci-fi.

Still, I need to make this caveat crystal clear: Even if you subscribe to the Many-Worlds Interpretation, neither Everett nor anyone else has ever proposed a physical mechanism for traveling between universes. Speculate all you want, but you're stuck here.

Level 4 Multiverse: If a Universe Is Mathematically Self-consistent, It Exists

At Level 4, things get even stranger. From Levels 1 to 3, there is the assumption that the laws of physics have at least some passing resemblance to those of our own universe. In the Level 4 multiverse, Tegmark supposes, "All structures that exist mathematically also exist physically," though it's not even clear how many mathematically describable universes there are.

For all we know, it's possible that there's a universe out there with only one of our fundamental forces or none. Because we haven't solved all of physics in our own part of the multiverse, even if Level 4 exists, we can't say with any certainty what those universes are like.

Part of the problem that we're encountering throughout this chapter centers on the idea that we really don't know whether the parameters that describe our universe are absolutely necessary and hardwired into the requirement that the universe be self-consistent or whether they're completely arbitrary. It's entirely possible that Tegmark's Level 4 multiverse predicts an infinite number of universes . . . or only one.

And if you're already suffering from a headache from the dizzying array of multiverses, thinking of ultimate ensembles is unlikely to help.

But really, we just want to focus on Level 1 and 2 multiverses. After all (in case you've forgotten), the whole point of this discussion is to address the question of whether the laws of physics vary throughout the universe.

IS THE UNIVERSE JUST RIGHT FOR US?

I've made this disclaimer already, but I wanted to emphasize it: While symmetries are going to give us insight into the nature and form of physical laws, they aren't going to tell us *anything* about the specific numbers that go into the laws. We're not going to (or at least we haven't yet been able to) "derive" the mass of an electron. There may be something fundamental about the universe that will allow us to derive all of the physical constants, but for the moment, we're totally in the dark. What that means is that we don't know whether the physical constants are hardwired into the laws, or whether, like the temperature outside on a particular day, they're simply something that happens. Symmetry tells us how to write down the equations but doesn't tell us how to assign the variables.

There are a lot of parameters (the charge of the electron, for example) that seem to be put in more or less by hand. Perhaps these parameters really can vary over an enormous universe, and it's only a particularly lucky region (our observable universe, for instance) that happens to be just right to house any sort of complex life at all.

There's nothing mysterious about the fact that we happen to live in a region where the laws are just right for human habitation. *Of course* they are! We wouldn't be here discussing them otherwise. That said, most physicists *really* don't like anthropic arguments. For most of us, there is a deep and abiding hope that at some point in the future, we'll be able to find a Theory of Everything based entirely on first principles.

Assuming that they're not hardwired, how finely tuned do the laws need to be for us to exist? What are the odds?

Let me anticipate a typical question about the fine-tuning of the universe. Why does light travel at 299,792,458 meters per second? The short answer, as we've seen, is that it's much more reasonable to simply say that light travels at 1 light-second per second and leave aside the question of defining a meter as a historical accident.

In other words, the values of parameters with units are almost always irrelevant because they clearly depend on what units you choose to use. I bring this up because there are a few ways of combining physical constants so that all of the constants drop out. In particular:

$$\alpha = \frac{e^2}{\hbar c}$$

is known as the fine structure constant (or FSC, for short) and it's a pure number without any units at all.

What is all of this? In the equation, e stands for the charge of an electron; c is, of course, the speed of light; $\hbar$ is known as the reduced Planck constant.* It shows up whenever quantum mechanics is involved. The fine structure constant has a value of about 1/137.03599908 and is one of the most exquisitely well measured quantities in all of physics. And while it may be well measured, we have absolutely no idea where it came from. This is way different from numbers in pure math. Pi, for instance, could be computed from first principles even if you've never seen a circle. As Richard Feynman put it:

* If you make casual reference to this at your next cocktail party (and you will!), be sure to call it *h-bar*. The pros will know what you're talking about.

> We know what kind of a dance to do experimentally to measure this number very accurately, but we don't know what kind of dance to do on the computer to make this number come out, without putting it in secretly!

The FSC is a measure of the strength of the electromagnetic force, and as you may notice, it's much smaller than one. In an objective sense, the electromagnetic force is actually fairly weak. On the other hand, in a relative sense, electromagnetism is immensely strong. Consider that the electrical repulsion between our sneakers and the floor are sufficient to easily overcome the gravitational pull of the entire earth.

There are at least twenty-five different dimensionless—and apparently independent—parameters involved in our standard models of cosmology and particle physics. Suppose we were to vary just this one, what would happen?

If, for example, the FSC were greater than approximately 0.1 (roughly 14 times its measured value), then carbon, and consequently anything heavier than carbon, couldn't be produced in stars. For carbon-based life-forms, that would be disastrous.

Or take another parameter, the strength of the strong nuclear force, the one that holds the nuclei of atoms together. If the strong coupling constant increased by only 4 percent, then protons would quickly bind to one another in the form of Helium-2, an isotope with no neutrons at all. Stars would quickly burn out, forming only inert helium, and nothing of interest would ever get created.

Most of the fundamental parameters seem to introduce exactly this sort of issue. We live in a universe where the parameters are tuned just right for our existence. This leaves us with a few possibilities, none of them terribly satisfying:

1. The universe was made for humans specifically or complex life generally.

2. The parameters of the universe are the natural consequence of some as yet unknown physical law, and we're just damn lucky that law allows for our existence.
3. The parameters vary across the multiverse, and by necessity, we're in one of the (perhaps very rare) regions that supports life.

Option 1 is just not about physics, and that's why I don't find it satisfying. Option 2 may, in fact, be the case, but physics has yet to discover a Theory of Everything. Until that time, there's not a lot we can say about it, which leaves me feeling deeply unsettled. So what about option 3?

Rather than ask what would happen if the FSC (or any other parameter) were to vary, we could address the observational question of *whether* it varies, and this means looking out into the vastness of space.

If we want to look into how the universe varies over cosmological distances, we have to start by observing objects that can be seen across billions of light-years. Fortunately, nature has provided us with the perfect lighthouse: quasars. Quasars are essentially giant black holes that absorb huge amounts of material. As the material falls in at speeds approaching the speed of light, it heats up and produces enough radiation to be seen across the universe.

The space between the quasars and us is filled with clouds of gas, and as the radiation heads toward us, the clouds absorb some of the incident radiation. The clouds absorb light only at particular wavelengths, and these wavelengths are determined by the value of the FSC. If you vary the FSC, those wavelengths will change as well.

Beginning in 1999, John Webb of the University of New South Wales and his collaborators decided to test whether the FSC varies over space and time, by looking at photons absorbed by various ions of iron and magnesium in very distant clouds. By measuring the relative wavelengths of the absorbed photons, they could compare the FSC at cosmological distances to what we measure in the lab here on earth.

What they found was extremely surprising. Observing distant galaxies in one region of the sky, the FSC seemed to have a value about 1 part in 100,000 *higher* than on earth, while in another part of the sky, the FSC seems to be about 1 part in 100,000 *lower* than on earth.

If the result is true, it's a very big deal. It means that somehow—and remember, we don't know where the value of the FSC comes from in the first place—the FSC varies throughout the universe. This flies in the face of the Cosmological Principle.

Two very important facts: The result is incredibly small, even if it's real. Nothing that Webb and his team observed would make either end of the observable universe unsuited for human life. You'd have to go much, much farther away than that. Second, most physicists don't currently buy the result. The signal is relatively tenuous and refuted by a number of other groups. I'm not going to take a big bottle of Wite-Out to my textbooks. If the laws of physics vary at all within our universe, they vary very, very little.

Anyway, here's the good bit. That variation still seems so trivial, if it exists at all, that we can introduce another symmetry:

> **Translational Symmetry:** The laws of physics are exactly the same at all places in the universe.

The large-scale homogeneity, the overall uniformity, in the structure of the universe reveals (or at least hints at) the underlying Translational Symmetry in the universe.

DYSON SPHERES AND AN INFINITE UNIVERSE

So, on the largest scales, there are no special places in the universe, and from that it follows that there is a Translational Symmetry in physical laws. That is the first part of the Cosmological Principle.

The second part of the Cosmological Principle *sounds* very similar, but there's a twist. Not only is the universe (roughly) the same in all places, but it looks more or less the same in all directions. The two don't have to go together, by the way. A honeycomb (or a Borg cube) is roughly the same no matter which cell you (assuming you to be a bee) are in. On the other hand, because of the hexagonal pattern of the cells, it looks different if you look toward a corner than if you look toward a side. That is, it's not isotropic.

But what about the universe? Is it more like a beach ball or a beehive? The best way to probe the universe at the largest scales is by looking at the cosmic microwave background (CMB). As I mentioned earlier, the CMB is the remnant radiation from when the universe was much, much younger than it is now.

This radiation is not exactly uniform. Some spots are a tiny bit hotter than the average, and some spots slightly cooler. These differences are the cosmological equivalent to static on an old-fashioned television. They represent seeds from random quantum fluctuations in the very early universe.

The differences are minuscule, however, only about 1 part in 100,000. Though the map looks highly regular, with splotches laid out

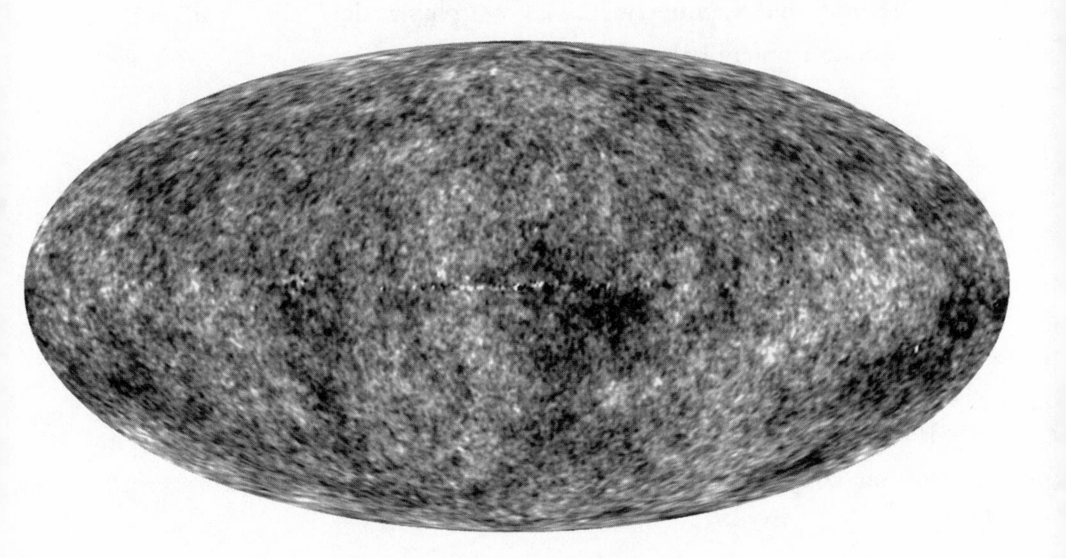

more or less at random, a number of research groups have done detailed searches to find directions with more structure than others.

A few such discrepancies have been found, subsequently dubbed the *Axis of Evil.* Taken at face value, an Axis of Evil would suggest that there really is a preferred axis to the universe.

By way of comparison, think about the earth. Our earth spins about an axis that passes through the North and South Poles, and because of the spin, the earth bulges out near the equator. A preferred direction can ruin the otherwise perfect symmetry of a sphere and turn it into a grotesque, asymmetric oblate (bluish in our case) spheroid.

Is the universe's Axis of Evil real, or just a statistical fluctuation? The current consensus seems to be that it's just randomly generated quantum noise. The problem is that, unlike experiments done in a lab, we have only the one universe, and because it evolves so slowly, we essentially get only a single snapshot.

The apparent isotropy of the universe, that it is essentially the same in all directions, suggests at least the possibility of yet another physical symmetry:

> **Rotational Symmetry:** The laws of physics don't change if you rotate a system in its entirety.

Which brings us back to the original question—hey, here we are!—of why the sky is dark at night. Again, it's not good enough to simply point out that the sun is on the other side of the earth. That's true, of course, but it's also true that the sun is not the only star in the universe. There are so many stars out there that when you crunch through the numbers, it's a wonder they don't fry us in an instant.

We've already seen that the universe is large—potentially infinite. If the universe really is more or less the same in all directions, then the farther away from earth you observe, the more stars you'll see. On the other hand, each of those stars will appear dimmer and dimmer, the farther away they are from us.

Which is more important: the virtually countless stars in the sky, each roughly the intrinsic brightness of our own sun, or the fact that each individual star is dim? To understand both of these effects, I'm going to invoke symmetry and, by way of illustration, offer a few hints on how to search for extraterrestrials.

In 1960, Princeton physicist and futurist Freeman Dyson proposed a way of searching for incredibly advanced civilizations. A truly advanced species might construct a giant sphere (known as Dyson spheres in sci-fi) around their star that would capture all of the incident radiation.

DYSON SPHERE

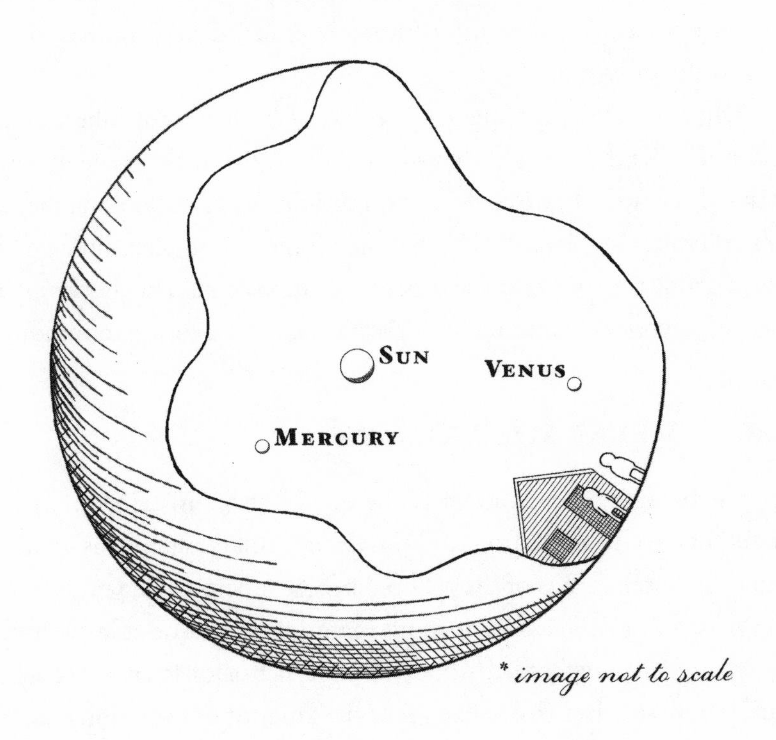

Radiation emanates from a star symmetrically—a function of both the symmetrical nature of the electromagnetic force and the fact that stars are nearly perfect spheres. The Dyson sphere captures the radiation uniformly on its surface, and if it's built the right distance from the parent star, the entire interior surface of the sphere becomes habitable.

To put things in perspective, if we were to build a Dyson sphere with a radius of approximately 1 Astronomical Unit (making the interior surface roughly room temperature), the available real estate would go up by a factor of a few billion compared to what we currently have available. Talk about a market crash! This would, however, allow for populations in the quintillions. Not to belabor the difficulties, but the impossibility of getting the raw materials to actually construct such a thing is likely to prove a major sticking point.

Dyson knew that all of the radiation that was absorbed would eventually be reradiated. The outside of the sphere would equilibrate at room temperature and would ultimately emit infrared radiation back out into the universe.

Dyson wasn't actually suggesting that we build a giant sphere around the sun. Rather, he thought that supercivilizations might come up with a similar idea, and if so, we could *find* supercivilizations by looking for giant infrared emitters. Given their awesome power, it is unclear how good of an idea contacting one of these supercivilizations would be, however. The image of a mosquito landing on a *Tyrannosaurus rex* comes to mind.

THE INVERSE SQUARE LAW

Suppose that instead of building a sphere at 1 Astronomical Unit, we were to build one twice as far from the sun. In our solar system this would be around the inner radius of the asteroid belt, just before you get to Jupiter.

We're going to need a lot more material. If you double the radius of a sphere, the surface area goes up by a factor of four. It's a square relation, you see. But this means that the amount of radiation reaching any point on the sphere goes down by a factor of four as well. The brightness of a source goes as the *inverse square.*

Dyson didn't invent the idea of an inverse square law. It has been known since antiquity, and shows up in all sorts of contexts. As a galaxy gets more and more distant, it appears to grow dimmer and dimmer as the inverse square of the distance.

Inverse square is also the rule for gravity. The more distant you are from the sun or the center of the earth, the weaker the gravitational force will be. This too becomes one-fourth as large as the distance doubles. The connection between gravity and the inverse square law formed the catalyst that ultimately prompted Isaac Newton to finally publish his *Principia Mathematica* in 1687. In it, he basically derived all of calculus and invented much of what we now know as physics, including his famous laws of motion. He may well have published all of this only as a consequence of a bet.

In 1684, Edmond Halley (the comet guy), Christopher Wren, and Robert Hooke—all eminent scientists of the time—were debating why it was that planets travel in elliptical orbits. Recall, Kepler had discovered this fact observationally 75 years earlier. Wren even offered a cash prize to whoever could solve it.

Halley was convinced (correctly) that planets were attracted to the sun with a force proportional to the inverse square of the distance, but he really couldn't get much further than that. So he went to Newton. While he hadn't yet written his masterwork, Newton was already a professor at Cambridge and generally acknowledged as a genius of the first order. As a contemporary described it:

> In 1684, Dr. Halley came to visit at Cambridge [and] after they had some time together the Dr. asked him what he thought the curve would be that would be described by the Planets supposing the force of attraction toward the Sun to be the reciprocal to the square of their distance from it.

> Sir Isaac replied immediately that it would be an ellipsis (an ellipse). The Doctor, struck with joy and amazement, asked him how he knew it. "Why," saith he, "I have calculated it."

The popular understanding is that Newton derived all of this some 25 years earlier while at home from school due to an outbreak of the plague. Whether this is true or not is unclear. What is clear is that he couldn't find his original papers, and so promised Halley to redo the calculation. Three years later, in 1687, he had published the *Principia*. Under the circumstances, it would have probably been ungentlemanly of Halley to collect on the bet.

The implications of the inverse square law are pretty significant. Remember, the law governs not just gravity but light as well, which is what got us talking about it in the first place. Imagine an infinite universe uniformly filled with galaxies. For ease of the argument, think of the galaxies as the ancients did with stars: embedded on celestial spheres. We'll change it up a bit by imagining many, many spheres. The more distant spheres, the more galaxies will fit on any given one.

Suppose one of these spheres is 10 million light-years from us, and a second sphere is 20 million light-years away. The galaxies in the smaller sphere will each appear four times as bright as their more distant counterparts. On the other hand, the distant sphere will have four times as many galaxies. Multiply it out, and both spheres pour the same amount of light onto earth. In an infinite universe, there are an infinite number of these spheres. Add them up and you get an infinite beam of brightness wherever you look.

If you suppose that astronomers are just overthinking things (or if you think I've cheated with my hand-waving argument), go to the middle of a forest. Nearby trees will look big. More distant trees will look small, but there are so many of them that if you're far enough into the woods, you won't be able to see out in any direction. Now suppose that those trees were on fire and had the surface brightness of the sun. The cosmologist Edward Harrison puts it rather poetically:

> In this inferno of intense heat, the Earth's atmosphere would vanish in minutes, its oceans boil away in hours, and the Earth itself evaporate in a few years. And yet, when we survey the heavens, we find the universe plunged in darkness.

This has come to be known as Olbers's paradox, so named because of the *last* person to describe it. Heinrich Olbers described the paradox that bears his name in 1823, but the idea is nearly as old as the Copernican principle itself, and goes back (at least) to 1605 and Johannes Kepler, who wrote of the terrifying possibility of an infinite stellar distribution in his *Astronomia nova*:

> This very cogitation carries with it I don't know what secret, hidden horror; indeed one finds oneself wandering in this immensity to which are denied limits and center and therefore all determinate places.

Kepler understood the implication of a symmetric universe, and he didn't like it. He thought that there must be a limit to the stars, and that they must be "enclosed and circumscribed by a wall or a vault."

We can't see infinitely far away because the starlight from objects farther than a few tens of billions of light-years hasn't yet had time to reach us. The vault that Kepler proposed is the beginning of time.

One symmetry, the Rotational Symmetry of physics, gives us the inverse square law. That and another symmetry, the homogeneity of the universe, combine to give us the mystery of why the sky is dark at night. That mystery is solved by an asymmetry: time.

While we've already seen that the *flow* of time seems very symmetric, there is clearly something asymmetric about time as a coordinate in our universe: It has a beginning. The beginning of the universe is the source of all of our problems when trying to figure out why the entropy in the universe was (and still is) so low. It's also what allows us to sleep at night. But wait a second . . .

WHY ARE PAST, PRESENT, AND FUTURE OUR ONLY OPTIONS?

We have been making a big assumption. We've been talking about the symmetries of space as though it were obvious that we live in a universe with three dimensions of space and one dimension of time. Everything we know about the standard model is built on the assumption that the universe has 3+1 dimensions (as the experts like to say), but it doesn't actually tell us *why* that has to be the case.

Not everyone takes a 3+1 dimensional universe for granted. One of the popular, if very speculative, approaches to understanding the unifying laws of the universe is known as M theory. M theory says, among much else, that the universe has ten dimensions in space and one in time. All but three of those spatial dimensions are presumably very small; they would essentially be a Pac-Man universe on scales not only much smaller than you and I, but also on scales smaller than atomic nuclei.

Suppose, for a moment, that M theory is right (though this is not the consensus view of physics), then it's possible that there's somewhere in the multiverse with more than three macro dimensions. However, the Anthropic Principle strongly suggests that nothing interesting could live there. Somebody is going to get up in arms and say that we don't *know* that life has to be like it is here on earth. That's true. I admit I'm only assuming, among other things, that complex molecules and atoms heavier than hydrogen need to be able to form. Because we've never seen extraterrestrial (let alone extrauniversal) life, I could be wrong. That's a chance I'm willing to take.

So what's wrong with anything but 3+1?

Edwin Abbott's *Flatland* is a narrative about a two-dimensional world that illuminates what we in our 3-D world can think about a 4-D world. A square tells the story. Yeah, a square. It's all about his civilization and physics. There may also be some political commentary thrown in there. I assure you that it's more interesting than it sounds.

The problem with such a world is one of complexity. To pick a gross example, imagine yourself as a two-dimensional, contiguous amoeba. A mouth-type opening takes in some food. How does your digestive system work? Well, presumably, there's a tube running through you, ending at your rear. The problem is that in 2-D, such a tube would split you in half. In other words, for your digestive system to work, your mouth would also have to serve double duty as your butt.

Grossness aside, there's a general problem in two dimensions, let alone in one. Systems and organisms simply can't be complex enough to form anything approaching intelligence.

It's easy to think about two-dimensional universes because we can draw them on paper or computer screens. It's far harder to visualize what life would be like in a universe with *more* than three dimensions. We have to at least consider the possibility, however. If M theory is right and there really are ten dimensions, why are so many of them compact and only three of them big?

We can say a fair amount about how physics works in a universe with more than three dimensions. We talked about the light-gathering power of a Dyson sphere as the radius gets larger and larger and concluded that the intensity of light drops off in proportion to inverse square of distance. The inverse square law isn't an accident. It's entirely a function of the fact that we live in a three-dimensional universe.

That inverse squares show up all over the physics of our universe is for the exact same reason. The intensity of a gravitational interaction falls off as the square of the distance between two stars. The intensity of the electromagnetic interaction falls off as the square of the distance between two protons, and so on.

In a higher number of dimensions, things get weird. For instance, if we lived in a four-dimensional universe, then we'd have an inverse cube law. If we lived in a five-dimensional one, we'd get an inverse fourth-power law, and so on.

This doesn't sound like a big deal until you realize that higher-dimensional universes (with their inverse cube or inverse-fourth gravity laws or whatever) don't have any stable orbits. In other words, in a four-dimensional universe, the earth would either spiral in toward the sun or fly away. We wouldn't get to enjoy the 5 billion or so years of nearly constant sunlight that we do in three dimensions.

This is true for all orbiting bodies (including planets, comets, stars in the galaxy, and so on), but our universe is dimensionally suited to life in another important way. Because electromagnetism (in our universe) also obeys an inverse square law, atoms wouldn't be stable in higher dimensions and would all spontaneously collapse. It's *really* hard to imagine complex life without atoms, and even tougher to imagine having this conversation without the existence of life.

It may occur to somebody reading this that electrons don't "orbit" atoms in the same way that planets do the sun. True enough, but if you grind through the equations in quantum mechanics and do the problem correctly, you hit the same complication. No stable atoms. Sorry.

So we're limited to three dimensions in space, but what about perhaps having more than one dimension of time?

MIT physicist Max Tegmark (who gave us our multiverse hierarchy) has a very nice discussion of what life would be like in such a universe:

> If an observer is to be able to make any use of its self-awareness and information-processing abilities, the laws of physics must be such that it can make at least some predictions. . . . If this type of well-posed causality were absent, then not only would there be no reason for observers to be self-aware, but it would appear highly unlikely that information processing systems (such as computers and brains) could exist at all.

For us, with our paltry one dimension of time, the future and the past seem fairly unambiguous. It's as if you were walking down a narrow hallway. "Forward" in that case, seems straightforward.

But if you were standing in the middle of a large ballroom, you can go in any direction that you please. "Forward" and "backward" are no longer such simple concepts. It's the same with time. In a universe with two time dimensions (and at least two space dimensions), you can't say anything useful about the future.

One of the things that makes an intelligent observer is that I (assuming I am one) can look around and, based on the state of things around me, determine with some probability what will happen elsewhere at some point in the future. Where my arrow will land, that the animal I'm hunting will get tired eventually, and what happened to

Tuk-Tuk when he ate those colorful berries. With two dimensions of time, you simply can't do that. I'll skip the math, but the basic idea is that the future isn't very well defined if you've got two time dimensions. If there's no inference, there's no prediction and no science. It's very hard to imagine how creatures in those sorts of circumstances could make any decisions at all.

But perhaps a universe with two dimensions of time would be even more devastating for interpersonal relations. Every person (and particle, for that matter) would move through two different times, t_1 and t_2. But these two times can't proceed at exactly the same rate, because if they did, it would be exactly the same as if the universe had only one dimension of time. Perhaps if you were alone in this freaky universe, you probably wouldn't even notice that anything is wrong.

But things get awkward if you have a friend.*

Normally, if you meet up with someone, it's because you're at (more or less) the same coordinates in space during an overlapping period of time. The problem is that if the two people are moving through the different time coordinates at different rates, then even though they might remain in the same place, they *won't* remain in the same time.

Put more simply, while your personal clock may run normally, unless your loved ones are all moving through both coordinates of time you're destined never to see them again. Life is unpredictable enough with one dimension of time; two would just be ridiculous.

We seem to have one dimension of time and three of space. And that seems to be just perfect for us. Besides being able to predict the future and live in stable orbits, our dimensionality, coupled with symmetry, a fixed speed of light, and the fact that the universe has a beginning also adds up to night being dark enough not to immediately evaporate. That works for me.

---

* Use your imagination, if necessary.

Chapter 4

EMMY NOETHER

IN WHICH WE DETERMINE WHAT SYMMETRY REALLY MEANS

Suppose you and your friends wanted to form a Fantasy Physics League—you know, the way kids do. Or indoor kids at any rate. Who would you want for your roster? Unless you're intentionally being contrarian, you'd probably go with Einstein. Everyone does.* You might name Max Planck, Werner Heisenberg, Erwin Schrödinger, or Wolfgang Pauli, the sort of big money players who have things named after them. (Check the index.)

Some overly sensitive person (me) might, at this point, have the gall to note that most of your roster, perhaps all of it, consists entirely of dead white men. After a few minutes of nervous hemming and hawing, you may finally remember Marie Curie, whom you'll recall as the discoverer of radioactivity and the first double Nobel laureate, a distinction

* In my first book, my co-author and I made just such a list and, no surprise, Einstein was at the top. Though my thoughts about this have evolved somewhat since then, if you're curious, the rest of our top five were Richard Feynman, Niels Bohr, P. A. M. Dirac, and Werner Heisenberg.

unique to her for over half a century. The point is, it's not exactly generous of you to call her in off the bench; she should be a starter.

In fact, this chapter isn't about her, and it's a good thing too because, as one of the indoor kids myself, I've exhausted my supply of football analogies. Or basketball. Whatever.

This chapter, and, quite frankly, much of symmetry itself, centers on my own pick: the mathematician Amalie "Emmy" Noether. Most science hobbyists and even many physics students are not familiar with this giant of our modern scientific age. And that's a shame because few people in the twentieth century did more to explain how the universe ultimately works. Emmy Noether, and her major contribution,

Noether's Theorem, finally gives us the critical insight into why symmetry is so important.

EMMY NOETHER THREATENS TO OVERTHROW ALL ACADEMIC ORDER

In many respects, Noether's story parallels that of Einstein's. They were both born to Jewish families in the late nineteenth century in what is now Germany. He, in Ulm, Wurttemberg; she, in Erlangen, Bavaria. Emmy Noether's father was an eminent mathematician at the University of Erlangen, and Noether decided to follow in his footsteps.

This was no simple task. German universities in 1900 generally did not allow women either to enroll in classes (a policy that was enthusiastically enforced by much of the faculty) or to sit for examination. In 1898, the faculty senate of Erlangen went so far as to claim that admitting women would "overthrow all academic order." A far easier route would have been the one that was offered to her after high school. Noether had a proficiency in languages and could easily have moved into teaching English and French.

Instead, Noether opted to pursue her entire undergraduate education in mathematics by auditing classes and was still able to pass the graduation exams at a *Gymnasium* in Nuremberg in 1903.

In 1904, Noether began her doctoral studies at Erlangen because the ban on women had finally been lifted. Her thesis adviser was Paul Gordan, a very close collaborator of her father, Max. Like many other ostensibly pure mathematics of the era, Gordan's work found its way into the newly developing quantum mechanics, in this case into the so-called Clebsch–Gordan coefficients, which are used to describe spin and orbital motion of electrons.

Noether earned her Ph.D. in 1908 and had an extremely tough time finding an official academic appointment, despite her obvious gifts. Einstein, famously, faced similar troubles and was exiled to a Swiss patent office until after becoming world famous during his 1905 Miracle Year. Meanwhile, Noether spent the next 8 years as an unpaid researcher at the University of Erlangen, occasionally substituting for Max Noether in his lectures.

Emmy Noether was an expert in mathematical *invariants*. Because this is the first time we're encountering invariants and because they're *very* important in understanding symmetry, here's the boldfaced definition, the first definition that isn't a symmetry:

Invariant: A number that is unchanged by a transformation.

A *transformation* is something like a rotation or moving a system from one place to another. *Invariants* are the counterpoint to symmetries. While a symmetry describes the sort of transformations that you can apply to a system without changing it, an invariant is the thing itself that is unaltered.

Just to mess with your mind a little bit, let me give you an example of something that turns out *not* to be invariant under certain types of transformations: duration. Take something as basic as the ticking of a watch, the beating of your heart, or the orbit of the earth around the sun. There may be a psychological aspect to how fast time *seems* to pass, but most of us accept on a rational level that there should probably be some sort of absolute measure of how much time passes between two events.

Not so.

One of the strangest implications of Special Relativity, as we'll see in the next chapter, is that the amount of time between any two events is decidedly a function of the person doing the measuring. As a classic example, the pilot of a spaceship flying by at close to the speed of light will appear to age more slowly than normal. Attach a stethoscope to a monitor and you could measure his heartbeat on the go. If he flies past at 99 percent the speed of light, his onboard heart monitor might measure 100 beats per minute, but by your external measurement, it would only be about *2* beats per minute.

Nothing changed in this measurement except your perspective, and yet something very significant changed about the value that you measured. As the pros would say, "Duration is not an invariant of your state of motion." Because we normally travel at a very tiny fraction of the speed of light, we normally can't measure this effect at all.

But there are *lots* of invariant quantities. The strength of gravity, as we've seen, is inversely proportional to the square of the distance between the two bodies. But the magnitude of the force is completely *in*dependent of direction. You weigh the same in Canberra as you do in Kansas.

Noether wrote her Ph.D. thesis on invariants and continued doing

Klein Bottle

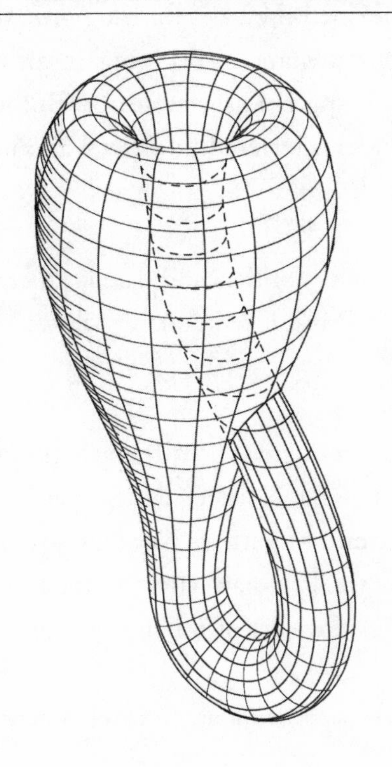

research on the topic during her subsequent time at Erlangen. If you're starting to get a picture of why Noether might be just the person to understand how symmetry *really* works in physical laws, you're not alone.

In 1915, Einstein published his Theory of General Relativity. It was, in short, one of the most revolutionary ideas ever published and almost instantly transformed our understanding of how space, time, and gravity really work. There was something incredibly elegant and deeply symmetric about relativity, but nobody really understood how it all fit together. The eminent mathematicians David Hilbert and Felix Klein invited Noether to the University of Göttingen in 1915 to help in unraveling some of the hidden symmetries.

Noether had met Hilbert before. In 1903, after completing her undergraduate studies, Noether spent a year in Göttingen and took classes with Hilbert and Klein, along with Karl Schwarzschild, who derived the first working model of a black hole, and Hermann Minkowski, whose mathematics formed the basis for the *special* theory of relativity.

Under normal circumstances, bringing in an expert at her level would have justified hiring her as a professor. But just as at Erlangen, biases against her gender interfered. Hilbert was outraged. At a faculty meeting, he exclaimed:

> I do not see that the sex of the candidate is an argument against her admission as a Privatdozent.* After all, we are a university, not a bathhouse.

Hilbert and Noether skirted the rules by listing Hilbert as a course instructor and then having Noether as the perennial guest lecturer, though this didn't extend to getting Noether any sort of paycheck. It wasn't until 1922 that the Prussian Minister for Science, Art and Public Education gave her any sort of official title or pay at all, and even

* Roughly equivalent to an associate professor in the American system.

then only a pittance. There is some evidence that the lack of anything more substantial was due not only to her gender but to the fact that she was a Jew, a liberal, and a pacifist. As Hilbert described it in his memorial address for Emmy Noether:

> When I was called permanently to Göttingen in 1930, I earnestly tried to obtain from the Ministerium a better position for her, because I was ashamed to occupy such a preferred position beside her whom I knew to be my superior as a mathematician in many respects. I did not succeed. . . . Tradition, prejudice, external considerations, weighted the balance against her scientific merits and scientific greatness, by that time denied by no one.

In all events, bringing her to Göttingen turned out to be an incredibly good idea. Almost immediately upon her arrival, Noether derived what's become known as Noether's Theorem and by 1918 had cleaned it up enough for public consumption. And this is where we pick up the physics part of the story.

Up until this point in the book, I've simply been telling you about symmetries and delving into their implications. You may have been thinking to yourself that there must be some sort of unifying principle to them. Right you are!

WITHOUT ANY FURTHER ADO . . . NOETHER'S THEOREM!

Noether provides the unification:

> **Noether's (first) Theorem:** Every symmetry corresponds to a conserved quantity.

If you are a bit underwhelmed, stand by. Conservation laws are the bread and butter of physics. We've seen conservation laws before, even if we didn't call them that. In every reaction we've ever discovered, positive charges and negative ones are always created or destroyed in perfect concert. If the Big Bang produced an electrically neutral universe, a very reasonable assumption, then conservation would demand that it still must be neutral today.

The relation between conservation laws and invariants is subtle. With an invariant, you take a system and do something like turn the axes or move the origin or move the hands of your clock and show that certain numbers, the invariants themselves, don't change. A conservation law, on the other hand, describes a quantity that stays the same over time. The total amount of energy—or charge—in the universe is a conserved quantity.

Despite the word *same* appearing in both descriptions, it isn't obvious just by looking that invariants and conservation laws should have anything to do with one another.

Here's a hint. As I noted in the first chapter, except for the weak force, there is no distinction between matter and antimatter. That is to say there is a symmetry, the C Symmetry (charge-reversal), between matter and antimatter. Therefore we expect matter and antimatter to be created and destroyed in equal numbers. Because matter and antimatter have opposite charge, the total charge in the universe needs to balance from moment to moment. This is a conservation law.

What Noether proposed sounds quite simple, almost content-free, until you get into the nitty-gritty. The conceptual problem for us is that Noether was a mathematician, which means that the details involve a lot of equations. And because we're not going to derive equations, it couldn't hurt to start off by giving you a few results, just so you know what to expect. Noether's Theorem predicts:

Time Invariance → Conservation of Energy
Spatial Invariance → Conservation of Momentum
Rotational Invariance → Conservation of Angular Momentum

This is a lot to digest. Each line describes a symmetry that we've already seen in the real universe. What I really mean is that the laws of physics don't change if you adjust the clock of the universe or move to a different place or point in a different direction. The arrows mean that if you have the first, the second necessarily follows.

We've made a lot of the fact that the laws of physics seem to be unchanging over time. This is more than just an assumption; there's some pretty strong evidence, including the Oklo site in Gabon. Noether's great contribution to physics is the mathematical proof that so long as the laws of physics don't change with time, then energy can't be created or destroyed.

And again, given the laws of physics really do seem to be the same everywhere, Noether's Theorem immediately tells us that there's a conservation of momentum. If you're flying through deep space, you can't expect to glide to a halt; you'll just keep drifting at the same speed forever. You may also know this as Newton's First Law of Motion.

It is incredible how little attention Noether's Theorem gets, even among people who work in and teach physics. As Lee Smolin has put it:

> The connection between symmetries and conservation laws is one of the great discoveries of twentieth century physics. But I think very few non-experts will have heard either of it or its maker—Emmy Noether, a great German mathematician. But it is as essential to twentieth century physics as famous ideas like the impossibility of exceeding the speed of light.
>
> . . . I've explained it every time I've taught introductory physics. But no textbook at this level mentions it. And without it one does not really understand why the world is such that riding a bicycle is safe.*

* In case you're missing the connection between Noether and riding a bicycle, it's all about the conservation of angular momentum. As a practical matter,

Now we get to the fun part. This is where we find out where the laws of physics really come from.

FERMAT'S PRINCIPLE

All of this jibber-jabber about invariants, symmetries, and conservation laws may seem a little abstract. So let's be more concrete.

Think of a comparatively simple system like using a slingshot to launch angry birds through the air to knock over primitive lean-tos sheltering green pigs. You could, at any instant, determine the force from the slingshot on the bird, the air resistance, the gravity, and the comparable interactions with all of the other solid bodies in the environment. You could then calculate the velocity of the bird accordingly. You simply perform these calculations again and again and voilà! You get the motion of the bird.

This is how the physics in Angry Birds works, and if it's good enough for them—and Newton and Galileo—it should be good enough for us.

But in some ways, the Newtonian approach to physics does us a disservice. For one thing it doesn't immediately give you any feel for *why* certain systems behave the way they do.

Let me ask you a simple question: Why does light travel in a straight line? Newton had an answer. Light normally travels in a straight line because there are no forces acting upon it. This is the crux of his First Law of Motion:

> **Newton's First Law:** Every body persists in its state of being at rest or of moving uniformly straight forward, except insofar as it is compelled to change its state by force impressed.

---

conservation of angular momentum also means that the earth will orbit around the sun at a constant rate.

In the first century BCE, the mathematician and engineer Hero of Alexandria suggested a different way of thinking about the motion of light. Light knows* where it wants to go. It simply picks the fastest route to get there. Hero played this game with mirrors and showed that if you reflected a beam off a mirror to a particular point, then no matter what point you pick, the shortest route will be the one that the light beam actually takes.

As it happens, Hero's solution independently produces the Law of Reflection: The angle of incidence equals the angle of reflection. If you want to try to test this rule for yourself, you can use a tennis ball instead of a photon and simply throw it at an angle against a flat wall.

Sixteen centuries later, Pierre de Fermat† generalized this for the motion of light under all circumstances:

> **Fermat's Principle:** Light takes the minimum amount of travel time possible.

It sounds, on the face of it, fairly simplistic, but it's really quite devious. How does light *know* how to take the quickest route? If you'd like a real-world example, go into the shallow end of a pool, and stick your leg into the water at an angle. You'll notice that your leg appears to bend downward at the surface of the water.

The Dutch physicist Christiaan Huygens, provided an entirely Newtonian explanation for the phenomenon. Light travels more

---

* Yes, I'm going to anthropomorphize photons. It's a metaphor. Go with it.

† The same guy who proposed that there was no integer solution to $a^n + b^n = c^n$ for $n > 2$ and for a, b, and $c \neq 0$. You may recall that he wrote an enigmatic, almost taunting comment in the margins, saying, "I have discovered a truly marvelous proof of this, which this margin is too narrow to contain." He was probably full of it. A full proof to Fermat's Last Theorem wasn't found until 1994 when it was discovered by Andrew Wiles. He had to develop almost entirely new branches of mathematics to do it.

quickly through air than it does through water. If you sat down to think about this, you might be surprised. After all, doesn't light always travel at *c*? Actually, the speed of light is really the speed of light *in a vacuum*. Once you start propagating waves through matter, everything gets slowed down. Roughly speaking, the denser the medium, the slower light will travel.

Huygens thought of light—correctly, as it happens—as a series of propagating waves. Huygens's Principle asserts that at every step in its propagation, light can be thought of as a series of outgoing circular waves. On the water side of the interface, the waves propagate more slowly and, as a result, interfere in just such a way as to make the light appear to bend downward.

Light doesn't intuitively seem like a wave in the way that, say, sound does. Sound waves have no problems bending around corners, but you can't *see* around corners.

Rather than strain ourselves trying to imagine the interference patterns of countless circular waves as they propagate across an interface, let's consider a much simpler (and much, much hokier) visualization of this.

Imagine a squadron of soldiers walking in perfect lockstep along a beach toward the ocean. The front row is at an angle to the shore, so one

LIGHT AT AN INTERFACE

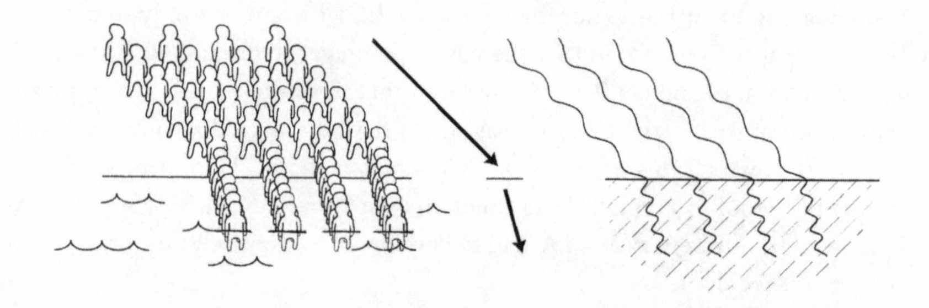

of the soldiers will hit the water first, and then another, and then another, until finally, the soldier at the far side of the first row enters the water. The same thing happens in every row, of course.

It's much slower going walking through water than on the beach, and so the soldiers who reach the water first will be significantly stalled. Others will hit the water shortly thereafter, and will nearly catch up in the interim. In the process, the angle will become far shallower.

Fermat's Principle gives us another approach, and one that doesn't depend on worrying about what happens on a moment-by-moment basis either to the beam of light or to the soldiers marching in a line.

Instead, consider a Hasselhovian figure sitting on the beach. He sees a swimmer struggling in the water. What route should he take? As before, he can run much faster on the beach than he can swim in the ocean. It makes sense for him to cover most of the horizontal distance on the beach and, once he hits the water, to make a nearly straight shot to the drowning swimmer.

The amazing thing is that if you crank through the math, Huygens and Fermat produce exactly the same path. Huygens's calculation doesn't say anything about the global properties of the system. Each soldier is walking in a straight line as best he is able, but because of the drag of the water, the *pattern* of the front row gets bent.

The empirical relation describing the bending of light as it crosses an interface had been known for a while. It is known as Snell's Law, which had been discovered* by the awesomely named Willebrord Snellius in 1621.

Snell's Law is a simple relation between the angle that light hits an interface, the relative speed of light in the different materials, and what angle the light comes out at when it emerges from the other side. In a practical sense, it's all you need to know to start making lenses or doing

* Rediscovered, technically. It had previously been found by Persian physicists over 600 years earlier and had been found by a number of others in the interim.

any other sort of optics. None of this wave interference or minimizing light travel time is even necessary!

So why is Fermat's approach, which is so circuitous, particularly useful? At best, you'll simply get the same result you would have gotten had you applied Huygens's Principle. Fermat's Principle isn't about getting the right answer; it's far more about understanding what's really going on.

Fermat's Principle has proven surprisingly robust to modern discoveries. One of the amazing predictions of General Relativity is that time runs more slowly near a massive body—like a star or a black hole—than when far away. Because light still wants to take the fastest route possible, it will go out of its way to ignore the gravitationally induced congestion arising from passing too near to a black hole. Put simply, the same principle that predicts the behavior of prisms also predicts that massive bodies will bend light.

The shortest route sometimes isn't a straight line. Suppose you want to fly to Beijing from Philadelphia. What route should you take? If you've

ever taken a transoceanic flight and kept yourself sane during the trip by watching the computerized map of your flight, you may have noticed that the plane does not take what you might naively think of as a straight line. Philly and Beijing are at roughly the same latitude, about 40 degrees. And yet, the path that you actually fly will take you north of Alaska.

This path, the great-circle route, does exactly the sort of thing that Fermat is suggesting. It's the route between Philadelphia and Beijing that minimizes the travel distance. You could find much the same thing by simply taking a globe, putting pushpins in the start and end points of your trip, and pulling a string between them until there's no slack. Despite the seemingly weird path on a map, a great-circle route is the most natural thing in the world. If you fly in a straight line and plot your path, Indiana Jones–style, on a map, you're going to cover a great circle without even trying.

But ultimately, as we'll see, Fermat didn't just tell us the fastest way to fly to Beijing; he also provided the groundwork for formulating all of physics on the basis of symmetry—a path that will lead us straight back to Emmy Noether.

HOW TO BUILD A BETTER ROLLER COASTER

Fermat's Principle is all well and good if we're worried just about the motion of light, but if we're trying to understand all that there is, we're going to have to dig a bit deeper. Fermat's Principle can't describe the creation of particles and antiparticles, the existence of forces, fields, the expanding universe, or about a million other observable facts of the universe we live in.

For mathematicians and physicists, and the distinction wasn't a terribly important one in the seventeenth century, these minimization problems found expression in trying to apply Newton's newly discovered laws of motion in new and interesting ways. One of the most famous problems of the day was the search for the Brachistochrone* curve.

Suppose you wanted to build a totally kick-ass roller coaster. You can make your tracks as frictionless as is humanly possible, but in trying to figure out the ideal shape, you come upon a little mathematical mystery. You'd like to design the roller coaster in such a way that a car starting at rest at one point will reach a fixed bottom as quickly as possible.

Though the puzzle of the Brachistochrone had been around for some time, in 1696 Johann Bernoulli—one of a very famous family full of outstanding mathematicians—claimed to have solved the problem and posed it, rather pompously, as a challenge to other mathematicians:

> I, Johann Bernoulli, address the most brilliant mathematicians in the world. Nothing is more attractive to intelligent people than an honest, challenging problem, whose possible solution will bestow fame and remain as a lasting monument. . . . If someone communicates to me the solution of the proposed problem, I shall publicly declare him worthy of praise.†

Bernoulli himself devised a particularly clever solution to the problem, justifying his self-praise, I suppose. Just as the roller coaster car will speed up as it gets closer to the earth, Bernoulli imagined building a very complicated lens, one made of thicker material up top and thinner below, so that light shining through it would go faster and faster, the closer it got to the earth. Then he simply applied Fermat's Principle and demanded that the beam travel the distance in the minimum amount of time.

---

* "Shortest time" in Greek, in case you're curious. And despite what you might guess, it has nothing to do with dinosaurs.

† Even *I* want to pants this guy.

Cycloid Curve

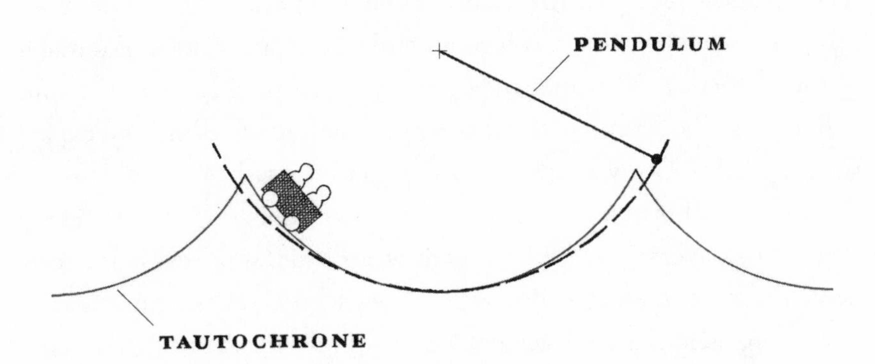

The solution was called an inverted cycloid, which looks very much like an ordinary bowl, but a very precise, mathematically derived one.

Much of this work was made possible by the fact that Isaac Newton had published his *Principia Mathematica* about a decade earlier.

By the time Bernoulli posted his challenge, Newton had a day job as master of the British Mint, but nevertheless he took a stab at the Brachistochrone problem. He solved the problem in an evening before going to bed, using a totally different geometric approach than Bernoulli. He was such a mathematical badass that he didn't even bother signing his name. Bernoulli figured it out anyway, remarking, "I know the lion by his claw."

It was big business to figure out the shapes of curves in the seventeenth century. Another famous challenge concerned the Tautochrone* curve. If a roller coaster is shaped like the Tautochrone, then no matter where you start the car, it will reach the bottom in the same amount of time. Christiaan Huygens—the same guy who described light in terms of waves—solved it in 1659. This is particularly impressive because it predates the *Principia Mathematica* by nearly 30 years.

---

* Literally, from the Greek, "same time."

I mention the Tautochrone because it turns out to have the exact same solution as the Brachistochrone: a cycloid. Besides its use as a mathematical problem, the Tautochrone is especially useful because, using the same principles, you could build a precision clock. For much of the history of the world, the only real precision clock was the sun, and because the seventeenth century was a period of sailing and exploration, solar clocks were decidedly not up to the task.

Near the bottom of the curve, you may notice that the Tautochrone looks very much like the arc of a pendulum. This is no accident. The whole reason pendulums make such good timekeeping devices is that, provided the displacement is relatively small, all oscillations take exactly the same time, which is why the arc of a pendulum fits so neatly into the bottom of a cycloid. Galileo noted this same fact, experimentally, as a young man. While bored and observing the chandelier in the cathedral in Pisa, he noted that the swing of the pendulum took the same amount of time (in heartbeats) regardless of how high it swung.

We're left with an interesting puzzle: The motions of particles and waves and light all seem to be governed by various minimization problems. Figure out the path that will get a light beam from point A to point B in the least amount of time and, lo and behold, you've found the actual path that it covers in the real world.

What problems like the Brachistochrone and Tautochrone show is that the same reasoning can be applied to the motion of massive particles as well. There's clearly something deep and important about minimization of travel time.

Newton and Bernoulli were geniuses, in part, because they were able to solve these problems without any general scheme for doing so. Essentially, they had to guess possible solutions until they came up with the shortest time.

That all changed in the eighteenth century when the mathematicians Leonhard Euler and his student Joseph Louis Lagrange found a general rule that allowed them to minimize the time, distance, or any other quantity along a trajectory.

We don't want to get bound up in mathematics, so let's think about a familiar example: our trip from Philadelphia to Beijing. There is literally an infinite number of possible paths to get from the one place to the other, though most of them involve jagged detours. We want to find a simple relation that connects the path we want (the one that takes a minimum amount of distance) with the realities of moving on a curved surface of the earth.

Euler and Lagrange described a way of calculating the shortest route, and it is not surprising that their method made ample use of Newton's newly discovered calculus.

THE UNIVERSE IS LAZY

All of this talk about minimizing time is fine if we're talking about light rays, airplane flights, and roller coasters, but it turns out that all of Newton's laws can be *derived* from similar assumptions.

In 1747, Pierre-Louis Moreau de Maupertuis devised what he referred to as the Principle of Least Action:*

> This is the principle of least Action, a principle so wise and so worthy of the supreme Being, and intrinsic to all natural phenomena; one observes it at work not only in every change, but also in every constancy that Nature exhibits. In the collision of bodies, motion is distributed such that the quantity of Action is as small as possible, given that the collision occurs. At equilibrium, the

* Though I'm convinced he invented the term so subsequent physics majors could complain about not getting action in college.

bodies are arranged such that, if they were to undergo a small movement, the quantity of Action would be smallest.

Action is one of those weird terms of art that sounds familiar until you realize that it's capitalized all the time and never replaced with a synonym. It does not mean what you think it means. It turns out that it doesn't even (quite) mean what Maupertuis thought it should mean.

To understand action, I first need to say a few words about energy. We've seen energy pop up time and again, and it doesn't always seem to refer to the same thing. Indeed, energy comes in lots of different flavors, the broadest categories of which are:

Rest Energy: The nuclear gas tank, able to be liberated at a rate of $E = mc^2$.

Kinetic Energy: The thing you think of when (and if) you think of "energy" and includes the energy of speeding bullets, locomotives, and flying Kryptonians.

Potential Energy: The energy of interaction; when you get winded climbing stairs, it's not because you're out of shape, it's because you're increasing your potential energy.

While energy can't be created or destroyed, it *can* be converted from one form to another. The tiny difference in rest energy between four hydrogen atoms and a helium atom produces an enormous amount of radiation. Jump out of a plane and your gravitational potential energy gets turned into a *lot* of kinetic energy.

Energy is at the center of all of physics, and in 1834, William R. Hamilton was able to put the work of Euler and Lagrange to good use. Just as Fermat showed that light wants to minimize the travel time, Hamilton's Principle, as it's come to be known, says particles will move about in a way that minimizes Maupertuis's Action, which also requires a slight definitional tweak in order to work.

The Action is, essentially, the average over time of a quantity known as the Lagrangian.* I know, I know, I'm giving you yet another technical term, but this one is especially useful. Take the energies of motion and subtract the energies of interaction and there it is:

Lagrangian = Kinetic energy - Potential energy

Send a toy rocket up into the air. It starts with a lot of kinetic energy, and when it reaches maximum height, all of the kinetic has turned into potential. On the way down, the potential turns back into kinetic.

In other words, on *average* the Lagrangian—the action—is zero. While the rocket is about as simple as you can get, there's an important lesson here: The action will be as small as possible if roughly half the energy is expended in motion and half the energy is contributed to potential. That's the goal: Minimize the action and get the actual trajectory that real rockets (or stars or atoms) will make.

The magic of Hamilton's Principle is that if you can figure out the Lagrangian for a system—the motion of a rocket, for example—you can then simply use Euler and Lagrange's approach to minimize the action and then get the trajectory of the rocket. In other words, if you know about the interaction energy in a system, and by extension the Lagrangian, then you know absolutely everything you need to predict its evolution into the future.

After all of this work, Hamilton's Principle gives you a way to go from a minimization principle to "derive" Newton's laws of motion. That is, in fact, what Hamilton's Principle was *designed* to do.

Hamilton's Principle seems like a really circuitous route to get to

* Which, despite the name, was actually found by Hamilton.

THE LAGRANGIAN OF A ROCKET

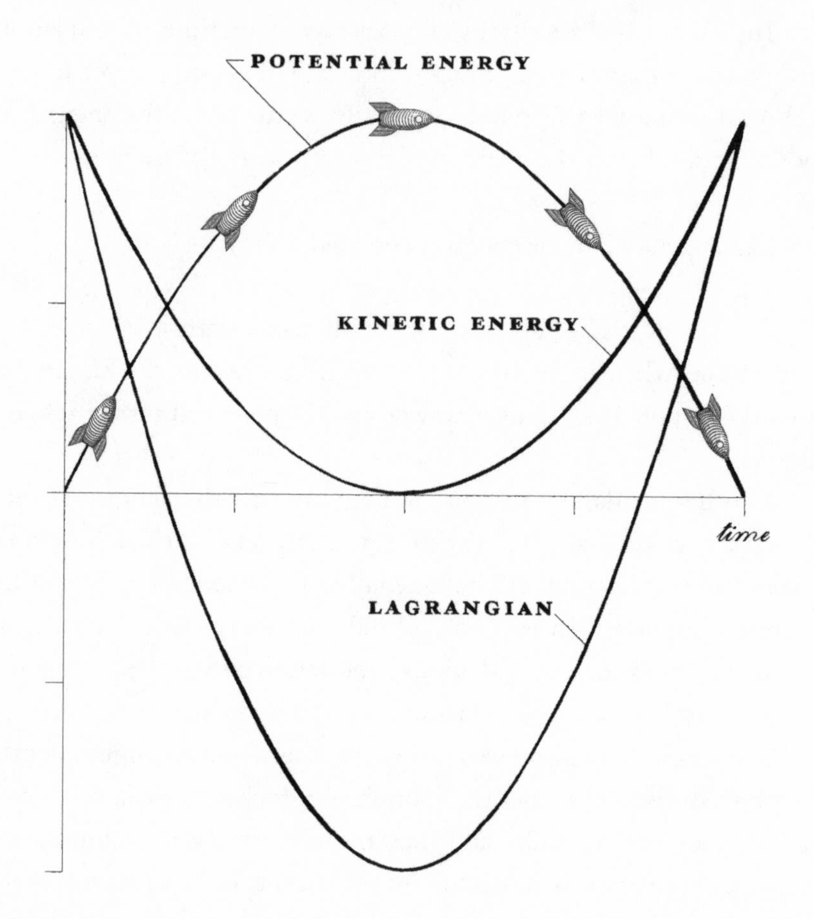

a result that we knew already. I made it all the way through my undergraduate mechanics classes without ever figuring out why we were doing what we were doing.

There are technical reasons, to be sure. Hamilton's Principle allows you to solve complicated problems with hard constraints on them that would have been essentially impossible if Newton tried to tackle them directly, but for our purposes, there's something even more important.

All of this discussion, from Fermat's Principle up to now, has been leading up to this point.

If there's a symmetry, it's the Lagrangian—essentially all of the energies involved—that remains invariant.

BACK TO NOETHER, AND WHAT HER THEOREM REALLY MEANS

I've listed a whole lot of symmetries throughout this book so far, but I never really addressed what the invariant *thing* is supposed to be. What is the thing that doesn't change if you look at it in a mirror, or change the clock of the universe, or run time backward, or rotate the entire apparatus?

Emmy Noether found the deeper truth. She realized that the Lagrangian is the thing that is symmetric.

Noether realized that certain manipulations change the energy, whereas others don't. For instance, if you scale up the entire universe, the distance between two bodies will increase, reducing their gravitational energy. But for those special symmetries, like rotation, that don't change the energy—and hence the Lagrangian—she showed that the symmetries immediately gave way to a conserved quantity.*

For instance, the laws of physics are the same today as they were yesterday, and therefore, according to Noether's Theorem, energy is conserved throughout the universe.

We've already seen that energy can change form. Chemical energy

* I need to make one little technical point, just to clear my conscience. Noether's Theorem specifically deals with so-called continuous symmetries. For displacement symmetry, the idea is that I could move my experiment by as small an amount as I want and nothing would change. Other symmetries are discrete, meaning that they are either/or. Charge conjugation is like this, as is reflection symmetry. *Either* you look at the universe in the mirror *or* you don't. There is no in-between.

is stored in food, which we can then eat and turn into heat (we *are* warm-blooded, after all) and motion. But if you were to add up all of the possible contributions in the universe—the $E = mc^2$ mass energy, the motion of all the particles, the gravitational and electrical interactions of all of those particles—you get one big number, and what Noether's Theorem tells us is that throughout all of time that number will remain exactly what it is today.

Noether didn't invent the idea of energy conservation. It is embodied in the First Law of Thermodynamics, but she did show that the first law was simply a consequence of the immutability of physical laws.

Likewise, the laws of physics are the same here as they are ten feet over there, so Noether's Theorem tells us that momentum is conserved. Conservation of linear momentum also wasn't new. It was discovered by Isaac Newton in the seventeenth century, and all three of his famous laws of motion are various ways of describing the conservation of momentum for an isolated system.

But here's the rub. We've already seen that time and space are closely related to one another. Relativity will show that you can exchange one for the other. As we will see, this means momentum and energy are simply two sides of the same coin.

Noether's Theorem describes much, much more. It describes and explains conservation of spin, electrical charge, of "color" (the equivalent of charge in the strong nuclear force), and on and on, ultimately providing the mathematical foundation for much of the standard model of particle physics.

Given all of that, it's a bit puzzling that Noether has, to some degree, been forgotten. Part of it, I think, stems from the fact that there are only a handful of calculations that one can do with Noether's Theorem. Once you've established that there is a conservation of energy or charge or maybe half a dozen other quantities, that's it. You don't need to or get to do any more useful calculations. Some of these conserved quantities might even have popped out of the equations using brute force, but who wants to do that?

We saw how Noether's story intersected that of Einstein's. Her life parallels Einstein's in other, sadder, ways. Like Einstein, she fled to the United States in 1933. Einstein settled in Princeton, at the newly built Institute for Advanced Study. Noether went to nearby Bryn Mawr College. And then only 2 years after coming to America, Emmy Noether was diagnosed with a cancerous tumor, and in the aftermath of a surgery, she died from infection. She was only 53. In Einstein's words:

> In the judgment of the most competent living mathematicians, Fräulein Noether was the most significant creative mathematical genius thus far produced since the higher education of women began.

Most important for our purposes, Noether finally explained why it is that symmetry shows up just about everywhere in the physical laws governing the universe. Symmetry isn't just something beautiful or elegant. The mere *existence* of symmetry ultimately gives rise to new laws! In a sense, she turned symmetry into order.

Chapter 5

RELATIVITY

IN WHICH WE FAIL TO BUILD AN INTERGALACTIC ANSIBLE

Admit it. One of the main allures of popular science is an enduring hope that you might use the knowledge to build a TARDIS or a warp drive. We've already touched briefly on multiverses, higher dimensions, and reversing the arrow of time. As another sci-fi staple, consider the ansible, introduced by Ursula K. Le Guin in her novella *Rocannon's World*. This is a device that, as she described it:

> Doesn't involve radio waves, or any form of energy. The principle it works, on the constant of simultaneity, is analogous in some ways to gravity. . . . What it does is produce a message at any two points simultaneously. Anywhere.

One of the funny things about this description is that it relies on one of the most popular misconceptions about gravity—that it travels instantly. Gravitational signals really travel at the speed of light. But that objection aside, are ansibles possible?

In a word: no.

I realize that I'm coming across *exactly* like that guy who keeps correcting your Klingon grammar, but that's too bad. The speed of light limit isn't just a strong recommendation; it's the law. Supposing you're persistent, your next question might be, *Why?*

When you were a kid, "because I said so" wasn't a sufficiently good reason to believe something, and you know what? It still isn't. To understand why, we're going to have to delve into—you guessed it—another fundamental symmetry of space and time.

WHY DOES IT *FEEL LIKE WE'RE* AT THE CENTER OF THE UNIVERSE?

Back in Chapter 3, we found, to nobody's surprise, that the earth is not at the center of the universe. We're traveling around the sun, the sun around the Milky Way Galaxy, and the galaxy around the local supercluster. When it's all added up, we're whizzing through space at a bit over a *million* miles an hour—I hadn't noticed. Did you?

Not that I want to give you yet another thing to worry about, but when you think about it, it *does* seem kind of terrifying that we're hurtling through space quite so quickly. It almost makes you want to spend all of your time indoors so as not to be swept off into deep space.

Galileo had the same sorts of existential worries nearly 400 years ago. He spent an enormous amount of effort fending off arguments from provincial types who *really* didn't like the idea of the earth going around the sun. In his *Dialogue Concerning the Two Chief World Systems*, Galileo goaded his detractors into playing with bugs and jumping around like fools just to prove his point:

> Shut yourself up with some friend in the largest room below decks of some large ship and there procure gnats, flies, and other such small winged creatures. . . . Then, the ship lying still, observe how those small winged animals fly with like velocity towards all

> parts of the room. . . . And casting anything toward your friend, you need not throw it with more force one way than another, provided the distances be equal; and leaping with your legs together, you will reach as far one way as another. Having observed all these particulars, though no man doubts that, so long as the vessel stands still, they ought to take place in this manner, make the ship move with what velocity you please, so long as the motion is uniform and not fluctuating this way and that. You will not be able to discern the least alteration in all the forenamed effects, nor can you gather by any of them whether the ship moves or stands still.

So long as your speed and direction are constant, there's no objective way to discern whether you or your bugs are moving or sailing. This is why, if you were so inclined, you could play pool or a game of catch aboard an airplane, and as long as there wasn't any turbulence, you would look exactly the same as you would look playing on the ground.

Even though you probably think "Einstein" when you think of relativity, Galileo got there first. Galilean Relativity contends that the only thing that matters is *relative* motion. You're traveling through space at a million miles an hour and so am I, but since we're moving together, we can comfortably construct a worldview around the idea that we're all sitting still.

This picture breaks down, at least from a practical view, when big things are involved. If you wake up from napping in the backseat during a road trip, you don't look out the window and assume you're standing still, with the mountains pulled toward one direction entirely for your benefit. That sort of self-centeredness is absurd, but if it makes you happy to pretend that you're sitting still and the rest of the universe is racing past you, that's A-OK with Galileo.

Practically, all that matters is the comparative motion of two bodies. If you and I are in spaceships heading toward one another at 1,000 mph, we'll experience the same sort of impact as though you were sitting there like a chump and I decided to ram into your spaceship at 2,000 mph. The velocities just add: 1 + 1 = 2.

Suppose that instead of hitting you head-on, I decide *not* to destroy my own ship, and instead, I shoot you with my laser cannon. The *L* in *laser* stands for light, which means that the individual photons travel at—are you ready?—the speed of light. But if Galileo is to be believed, then the speed of light will depend on the speed of the source. If I'm flying at you at 1,000 mph, you might reasonably expect that photons will fly at your ship at c + 1,000 mph, and not, as the name would suggest, at the actual speed of light.

What's the deal? Is there a single speed of light or isn't there?

THE AETHER AND THE OBLIGATORY MOCKING OF THE ANCIENTS

Up until the late nineteenth century, it seemed perfectly reasonable to suppose that if, for instance, Daleks were trying to ex-ter-min-ate you, then their lasers would travel faster if you ran toward them than if you (quite reasonably) ran away.

After all, Huygens showed that light is a wave, and all the waves studied until that time propagated through some sort of medium. Seismic waves (earthquakes) need to travel through rock. Water waves have water. Sound waves have air. Light, so the thinking went, must also need a medium. The existence of a medium is important because waves propagating through one will travel faster if they're going with the flow, than against it.

The working theory for most of the nineteenth century was that a thin fluid known as the luminiferous aether permeated the universe and that light waves were simply ripples in the aether. If the aether were real, the speed of light from a particular star would depend on the relative motion of observer and star as well as whether the aether itself was flowing.

I feel a little bit dirty right now. I'm following in the tradition of introducing the aether—an idea that probably wouldn't have occurred to you independently—and then immediately explaining why it turns out to be wrong. Then we all have a good laugh about how all previous scientific generations were dummies, unlike us.

Okay, the aether theory *is* wrong, but I'll say a few words about how we *know* that it's wrong. The argument is a lot simpler than the measurement. It's based on the fact that we're traveling around the sun, moving in different directions at different times of year. If the aether hypothesis were correct, the speed of light should depend on direction and should be a little bit different in the winter from in the summer.

Though a number of similar experiments were conducted both before and after, it's generally accepted that Albert Michelson and Edward Morley performed the definitive test of the aether hypothesis in 1887. Michelson and Morley used a device called an interferometer that could measure variations in the speed of light based on season, time of day, and orientation of the device. They found . . . nothing. No variation whatsoever. As the British physicist Sir James Jeans put it some 50 years later:

> The pattern of events was the same whether the world stood at rest in the supposed [a]ether, or had an [a]ether wind blowing through it at a million miles an hour. It began to look as though the supposed [a]ether was not very important in the scheme of things . . . and so might as well be abandoned.

If there isn't an aether, then it means that light isn't traveling *on* anything, which means that it's unlike any other wave that we've ever encountered. Like *so much else,* this too was explained by Einstein* in 1905. It is, of course, his Theory of Special Relativity.

HOW TO DESIGN A DEFLECTOR SHIELD

Einstein was fascinated with James Clerk Maxwell's unified laws of electromagnetism, and who could blame him? Maxwell's equations were a model of elegance and simplicity that described just about everything from why a compass needle points north to why a balloon will stick to a wall to the workings of atoms.

One of the equations, known as Faraday's Law, described how a time-varying magnetic field can generate an electric field. Another equation, Ampere's Law,† showed the converse: Time-varying electric fields can generate magnetic fields. This back-and-forth explains why physicists refer to electromagnetic waves when normal humans would say "light." Einstein realized that the breakdown of electromagnetism into electricity and magnetism is largely a matter of perspective.

Consider two spaceships in which the two captains (the two of us) are eyeing each other malevolently, thinking seriously about ramming into one another. A benign alien intelligence, Alice, wants to create a rudimentary deflector shield to protect us from our own worst impulses. The design is simple: She douses the hulls of our ships liberally with electrons. As our ships approach one another, the electrons push us apart.

This force shield works entirely through the power of electricity; magnetism doesn't enter the equation at all.

Having recognized the futility of trying to destroy one another,

* I mean, seriously, how awesome was that guy?

† You may have noticed that these laws are all named after other folks. That's why Maxwell's work is usually described as *unifying* the laws of electromagnetism rather than discovering them.

MAGNETISM IN A MOVING FRAME

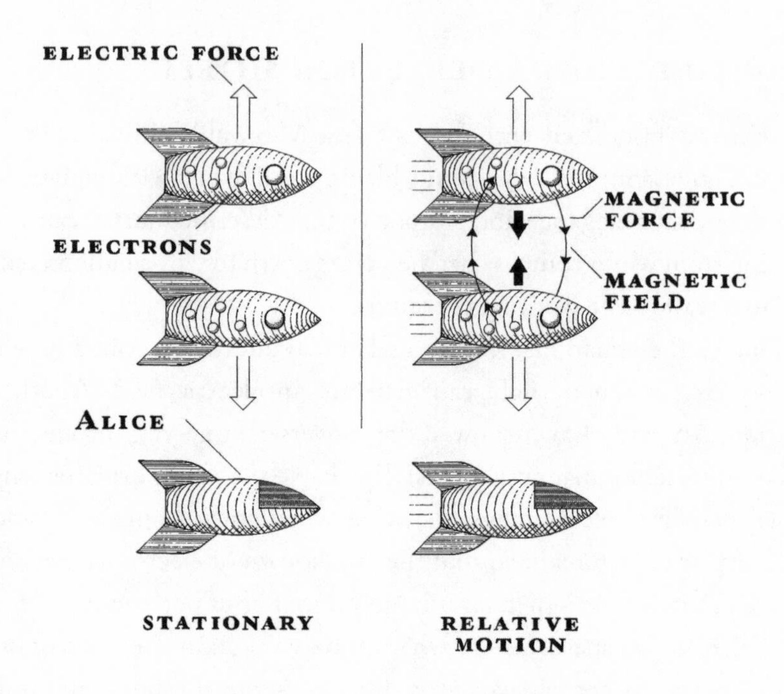

you and I decide to form an alliance and fly through space side by side in our electron-covered ships. Alice, who decides to stay behind, notices something surprising. From her perspective, electrical repulsion still pushes our ships apart, but the force isn't quite as great as when we were sitting still. Instead, each of our moving ships produces an electric current, and our currents generate a magnetic field. Because my ship feels the magnetic field generated by your ship and vice versa, the two ships magnetically *attract* one another.

To recap: Charge alone means repulsion. Charge plus current means some repulsion and some partially canceling attraction.

But here's where things become interesting. The only difference between the first case (electric repulsion only) and the second (electric repulsion and magnetic attraction) is whether or not you think the charges are moving. If Alice decided to exactly match our speed and heading, she could make the magnetic fields disappear entirely and, consequently, change the force between our ships without laying a single finger on us.

HOW EINSTEIN FIXED GALILEO

Einstein didn't see this as a problem; he embraced the strange behavior of electrical and magnetic fields and the constancy of the speed of light demonstrated by Michelson and Morley* and developed a new Theory of Special Relativity. He made two simple assumptions:

1. The same laws of electrodynamics and optics will be valid for all frames of reference for which the equations of mechanics hold good.
2. Light is always propagated in empty space with a definite velocity c which is independent of the state of motion of the emitting body.

Einstein assumed that there was yet another symmetry in the universe, and a pretty important one. Something—the speed of light and the laws of physics—remains invariant for observers moving at different velocities. Even though it's not written under his name, Einstein's postulates are subsumed under the heading of an explicit symmetry:

* Einstein made a number of contradictory remarks at various times regarding the level of influence that Michelson and Morley (whose work was 18 years old by the time Einstein had his Miracle Year) had on his second postulate of Special Relativity. At various times, he claimed to be completely unaware of their experiment, and at others, he claimed he knew of their work, but it didn't influence him. At any rate, without the experimental evidence provided by the Michelson–Morley experiment, Special Relativity would have been a hard case to make.

Lorentz Invariance: A law of physics will be written in such a way that the results are independent of the orientation or velocity of the system.

Einstein's first postulate, in particular, has come to mean that *all* experiments should work just as well as seen by all inertial observers, anyone who, depending on your point of view, is either at rest or moving at a constant speed without turning.

Just to be clear, there are a few caveats here. Not everyone gets to be an inertial observer. When you accelerate a car, for instance, you're pushed to the back of your seat and can plainly feel that you're not moving at a constant rate. It's this reliance on being an inertial observer that makes Special Relativity so special.

Einstein was onto something big, a symmetry we already saw in the context of the Second Law of Thermodynamics: Space and time aren't nearly as different as people had previously supposed. Depending on how you move, time and space can easily get mixed up with one another. We're not quite ready to get into the warping of space–time, but as a well-motivated warm-up, let's take a look at a surprisingly related idea, rotations in ordinary space.

THE PYTHAGOREAN THEOREM

No doubt you have come across the Pythagorean Theorem:

$$A^2 + B^2 = C^2$$

It's a deceptively simple equation. Variables *A* and *B* are the lengths of the two short sides of a right triangle, and *C* is the length of the long side, the hypotenuse.

The Pythagorean Theorem does far more than simply tell you about triangles for their own sake. It tells you how to compute the distances

PYTHAGOREAN THEOREM

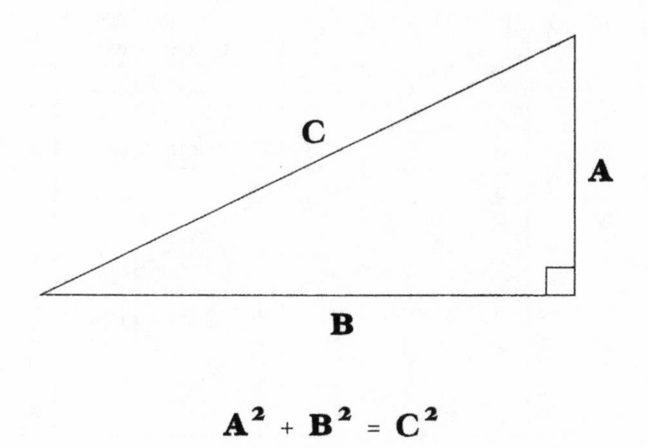

$$A^2 + B^2 = C^2$$

between points. You may remember problems like this from school: Walk 3 miles east and then 4 miles north. Plug through the calculation, and you'll find that you're 5 miles from where you started. To connect it to the real world, consider a small piece of a Washington, DC, transit map:

Many cities are conveniently designed so that streets run approximately along the cardinal directions of a compass. Washington, DC, is a perfect example: Numbered streets run north–south and lettered streets run east–west.

So, to pick an example found by an intensive search through Google Maps, if you wanted to walk from Judiciary Square station on the corner of 4th and E Street NW to the Chinatown metro station on 7th and G, you would start by walking approximately 600 meters due west (along E), and then 250 meters north (along 7th).

Of course, you could also simply take the Red Line subway, and if you plug through the numbers, you'd find that the subway trip is approximately 650 meters. It's just a practical application of the

D.C. Transit Map

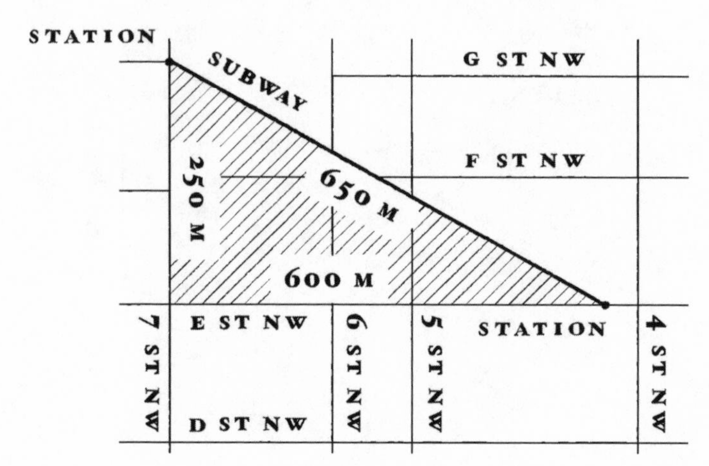

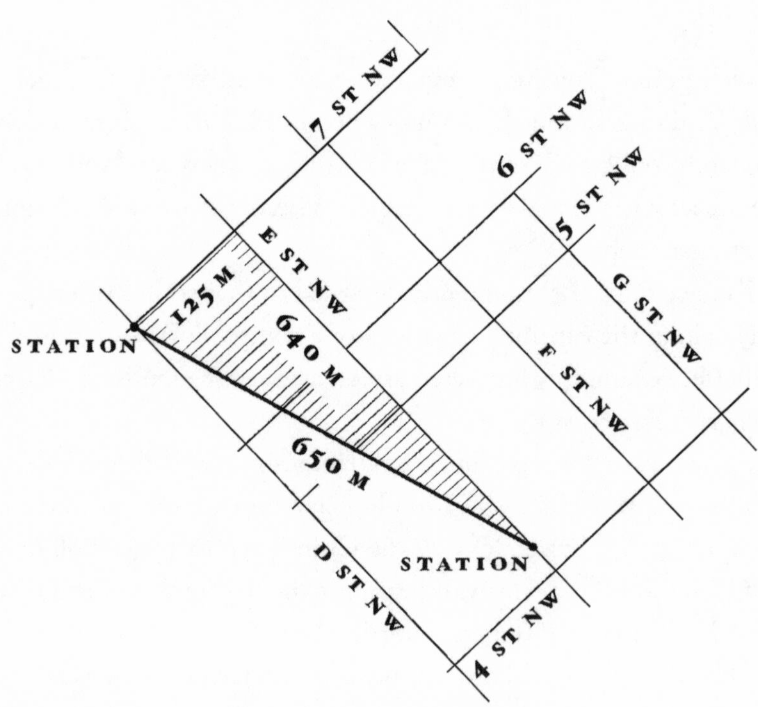

Pythagorean Theorem, albeit with the slightest relabeling of the numbers:

$$x^2 + y^2 = distance^2$$

We owe this convention to the work of the seventeenth-century mathematician and philosopher René Descartes. The Cartesian system imagines describing all of the events and objects in the universe on a sort of map. For instance, in the east–west direction, the convention is to label positions as x. In the north–south direction, we normally label positions as y. I'm going to ignore the possibility of moving vertically in an elevator, but if you were so inclined, you might label the direction of motion as z.

I'd be negligent if I didn't point out that Cartesian system breaks down on the surface of the earth. The earth, after all, is essentially spherical, which means that you can't make a perfect, undistorted flat map that covers the entire thing.* That's fine. That's why we're talking about something much smaller, like a few city blocks.

Suppose you stepped down into the Judiciary Square subway station and some particularly malevolent and efficient urban planner decided to come along and rotate all of the city streets while you were underground. He rips up and repaves all of the streets so that they are turned a few degrees from their original orientations. Why? Who can say?

The upshot is that while the subway stations don't actually move, their addresses do. The streets still make a grid, just a different grid. A

* As a fun side note, J. Richard Gott and I argued mathematically in 2007 that the map projection with the best combination of minimal distortions is the Winkel Tripel. Apparently, *National Geographic* had come to the same conclusion. They've been using the Winkel Tripel for their whole-world maps for more than a decade.

pedestrian in this new version of Washington still wants to walk from Judiciary Square to the Chinatown metro stop. She still walks down a lettered street and up a numbered one. Each leg of the trip is different from what it was before, and yet the subway trip that you take underground is *exactly* the same as it was before!

We've seen Rotational Symmetry a number of times already from the apparent isotropy of the large-scale universe down to the experimental fact that microscopic interactions really can't seem to distinguish one direction from another. We've even seen that Rotational Symmetry gives rise immediately to conservation of angular momentum. Long story short: You may have learned the Pythagorean Theorem as a kid, but it's anything but child's play.

WHAT DOES *DISTANCE* MEAN IN SPACE *AND* TIME?

Space and time are very similar to one another, but not identical. If Einstein's postulates of Special Relativity are correct—and to date, they have passed every experimental test thrown at them—then we're going to have to figure out a way of jamming space and time into a single "spacetime." Einstein himself warned of the danger of trying too hard to think in four dimensions:

> No man can visualize four dimensions, except mathematically. . . . I think in four dimensions, but only abstractly. The human mind can picture these dimensions no more than it can envisage electricity. Nevertheless, they are no less real than electro-magnetism, the force which controls our universe, within, and by which we have our being.

Let me put this in familiar terms, or at least familiar if you've memorized the entire *Star Trek* canon. The Vulcan home world is approximately 16 light-years from earth* and right now† Solkar, the great-grandfather of our own Mr. Spock, is a young spaceship pilot. Because we are separated from Solkar in space, but not in time, calculating the distance is easy: 16 light-years.

Separation in time, by itself, is also easy to quantify. You are now reading these words, and 10 seconds ago you were reading, "Let me put this in familiar terms." Assuming you are sitting perfectly still, these two events are separated in time by perhaps 10 seconds and not separated in space at all.

But what if events are separated in both space *and* time? If we were to point a ridiculously powerful telescope at Vulcan right now, we wouldn't see Solkar flying around in his spaceship. Instead, we'd see the events on Vulcan unfolding from 16 years ago. This is because we're seeing the signals traveling at the speed of light. Light signals will always have this one-to-one separation of space and time.

Einstein came up with his Theory of Special Relativity in 1905, but it wasn't until 1907 that the German mathematician Hermann Minkowski adapted Einstein's work to show how space and time really fit together in a way that would have made Pythagoras proud. As he put it rather grandiosely (but accurately):

> Henceforth space by itself, and time by itself, are doomed to fade away into mere shadows, and only a kind of union of the two [what we now call spacetime] will preserve an independent reality.

Minkowski realized that in some sense, space and time work in opposite directions. As a concrete example, consider Betelgeuse, the

---

* Gene Roddenberry, the creator of *Star Trek* himself, said that Vulcan was in orbit around the star 40 Eridani A, which is a fairly cool star, around 16.4 light-years from the sun. Please don't take my lunch money.

† We'll soon find out that "right now" is actually kind of ambiguous.

bright red star in the constellation Orion. There is some astronomical evidence that Betelgeuse might go supernova any day now.* Betelgeuse is about 600 light-years from earth, which means that even if we saw it blow itself up tomorrow, the actual explosion took place 600 years ago. We could describe the distance in space (600 light-years) or time (600 years), but combining the two is a bit trickier.

We have a hint, though. If the light from a supernova explosion is just reaching us, then it is, in a real sense, happening in the here and now. The speed limit of light prevented us from learning about it earlier. Minkowski created a variant of the Pythagorean Theorem by which time behaves *almost* exactly the same as distance, except for a minus sign:

$$distance^2 - time^2 = interval^2$$

This "interval"† is just a fancy way of combining distances in both space and time. It's also pretty clever. By definition, anything that we're just seeing now—regardless of how far away it is in space—has an interval of zero. More important for our purposes, the interval between two events is completely independent of your perspective. A fast-moving astronaut may measure a different distance between

* In astronomical terms, "any day now" means that it might go off in 100,000 years or more.

† The mathematically savvy may have noticed that for events that are separated more in time than in space, the interval squared turns out to be a negative number, making the interval an imaginary number. Don't worry about it. This is just a mathematical device to let us know that the two events are "time-like separated," which simply means that one event could affect the other. If the interval squared is positive, they're "space-like separated," which means that causality simply can't come into play. If math makes you want to rock yourself in a corner, please move on. There's nothing to see here.

two events and may measure a different time separation but will always agree with you on the interval.

We saw with the Pythagorean Theorem that it doesn't matter how you rotate the orientation of your streets, the distance between any two subway stations will always remain the same. The interval is exactly the same. Einstein said that all inertial observers should measure the same speed of light, which means that the interval between any two events should be the same no matter how fast you're traveling through space.

We're in deep now. Abstract thinking about symmetries ends up revealing surprising connections. From a mathematical perspective, rotating a coordinate axis reveals the exact same sort of symmetry in space as moving at different speeds does in spacetime. Both of these two transformations leave something invariant. For rotations, the distance between two points stays the same; for different speeds, "boosts" as relativists call them, the interval stays the same. Who could have seen *that* coming?

Imagine a Vulcan astronaut flying the earth-to-Betelgeuse route at a sizable fraction of the speed of light. As we'll see, he'll measure the total distance of the route to be less than 600 light-years. However, he'll *also* measure the delay in between Betelgeuse going kablooey and us measuring it as less than 600 years. All combined, he'll measure the same zero interval as we do, no matter how fast he's going.

HOW TIME GETS STRETCHED

Galileo made a totally understandable implicit assumption, one that seems almost too obvious to state aloud: Time runs at the same rate for everybody. And yet if you accept that light always travels at the same rate, a constant flow of time is a luxury that you can't afford.

Clearly time gets messed up, but so far, we haven't gotten any idea as to *how*. Suppose Solkar decided to buzz past the earth at half the speed of light. Provided he's moving in a constant speed and direction, he feels as though he is sitting still. That is one way of thinking about Einstein's first postulate of Special Relativity (The laws of physics are the same in all inertial frames of reference). As Solkar goes about his business—reading the paper, taking a nap, browsing the intergalactic net—he is, as far as he's concerned, moving through time and not through space.

On earth, on the other hand, we see him moving through both time *and* space. If Solkar settles down for what seems to us to be an 8-hour nap, by the time he wakes up, the ship will have traveled 4 light-hours.

The beauty of Minkowski's interval is that it's the same for all observers. There's a minus sign relating the spatial and temporal distances for a moving system, which means that they partially cancel.

In relativity, *everyone* can legitimately say that he or she is standing still. Solkar knows that time has passed during his nap, but because he feels as if he is standing still, he won't feel as though he traveled in space. In order for the interval to be the same measured on his ship as measured on earth, he must have slept for *less* than 8 hours. If you crunch the numbers, he's actually gotten only about 7 hours of sleep. Relativity can be even more disruptive to your circadian rhythm than daylight saving time.

Moving clocks run slow. This is not some trick of measurement; it's a real effect, albeit normally a very small one. To put things in perspective, even on the highest-speed Japanese bullet trains, time appears to run slower by only less than one part in a trillion. If a bullet train had been running since the beginning of time and suddenly came to a stop, we'd find that the inhabitants are somehow about 13½ hours younger than the rest of the universe.

The faster you go, the larger the effect. At 90 percent the speed of light, moving clocks run slow by a factor of 2.3. By the time we reach 99 percent the speed of light, things become crazy—clocks appear to be slowed down by a factor of 7! And understand, this isn't some weird optical illusion or mechanical effect because of the strain of the speed. *Everything* appears to be slowed down by the same factor. Solkar's heart, and all of his metabolic processes, would beat slower than normal; his computers would appear sluggish by normal standards; every single device capable of measuring time would appear to have slowed down to a crawl. And yet, from Solkar's perspective, everything appears to be running perfectly normally within the ship.

TIME DILATION

How much time is stretched for various things:

What?	How Fast?	Slowed by How Much?
Usain Bolt	23.4 mph	6 parts in 10 quadrillion
Fastest bullet trains	311 mph	1 part in 10 trillion
International Space Station	17,000 mph	1 part in 3 billion
The sun around the Milky Way	492,000 mph	1 part in 3.7 million
Half the speed of light	134,000,000 mph	1 part in 7
Muons in the atmosphere	99.5 percent of *c*	A factor of 10 times
Protons in the Large Hadron Collider (top speeds)	99.9999991 percent of *c*	A factor of 7,500 times

Although we can't actually build spaceships capable of moving at relativistic speeds, we *can* measure time dilation here on earth using particles called muons. A muon is almost identical to an electron—but 200 times heavier. As we've seen before, heavy particles, whenever possible, will decay into lighter ones, and the muon is no exception. After about 2 millionths of a second on average, a muon will decay into an electron and a neutrino-antineutrino pair.

Because muons decay so quickly, it's a wonder that they're around to be observed at all. Fortunately, the universe is nothing if not dedicated to the task of producing massive particles. When extremely high energy particles from space, cosmic rays, strike the upper atmosphere, a cascade of particles gets created, culminating in the production of muons. This means that the bulk of muons get produced more than 10 kilometers above the surface of the earth. This would be no big deal except for their incredibly short half-life. Even traveling at the speed of light, a typical muon could cover only about 600 meters before we'd expect it to give up the ghost. We'd reasonably suppose that virtually no muons should ever reach detectors on the surface of the earth. And yet, we *do* detect atmospheric muons all the time. We can even tell that they originate from space because we can see a big empty spot—a shadow of muons—where the moon is.

In 1941, Bruno Rossi and David B. Hall, both from the University of Chicago, measured the number of muons coming from the sky as measured at the top of a 2-kilometer-high mountain and at ground level. If Galileo were right, and time flowed the same for everybody, then virtually all of the muons should have decayed from top to bottom. Instead, based on the fraction of muons that did actually decay, Rossi and Hall estimated the muons' internal clock seemed to be slowed by roughly a factor of 5. They didn't decay in 2 millionths of a second but in 10 millionths of a second. The muons from space were hauling ass at roughly 98 percent the speed of light.

But relativity says far crazier things than moving clocks run slow. Einstein's first postulate of Special Relativity was that you can't ever tell if you're the one moving or standing still. It's easy to imagine things from Rossi and Hall's perspective. They're people, after all, and we tend to have an anthropocentric view of things.

But if you can manage a little particle empathy, put yourself in the muons' position. The muons also don't feel like they are moving at all. Here they are, newly born, and all of a sudden they see the ground—and Rossi and Hall—hurtling toward them at 98 percent the speed of light. The muons, provided they have the presence of mind to do the experiments, find that Rossi and Hall seem to be living in slow motion, by the same factor of 5 that we saw before.

Despite having studied relativity for a long time, this *still* seems crazy to me. How can it be that two people (or relativistic elementary particles, or whatever) view one another and believe that *they* are running normally while the other is running slowly? This simply seems to be logically inconsistent. And yet, it isn't.

Relativity affects space as well. Again, consider the view of Rossi and Hall's muons. From their perspective, it only takes about 1.3 millionths of a second from the 2-kilometer-high mountaintop to ground—a short enough interval that a good many of them will have survived the trip. But they—that is, the muons—can't possibly travel 2 kilometers in that time! Doing so would require the ground to race toward them at 5 times the speed of light.

Distance is just as screwed up by moving close to the speed of light, and by the same factor as time. Space just appears contracted along the direction of motion.

Einstein didn't exactly invent the equations of time dilation and length contraction. A number of physicists and mathematicians in the decade before the Miracle Year paved the road for Einstein's relativistic breakthrough. In particular, Hendrik Lorentz and George FitzGerald independently came up with a set of equations that described the stretching of space and time according to moving observers.

So why don't *they* get credit for discovering relativity?

They get shafted in part because Lorentz and FitzGerald discovered their famous transformations (still used today, by the way) as a way of explaining how the aether was, in fact, really a thing, despite the results of Michelson, Morley, and others. From Lorentz and FitzGerald's perspective, a spaceship would appear foreshortened not because of a fundamental change of space or time but because it and all of its measurement apparatuses were compressed at an atomic level due to their motion through the aether. It's not unlike the effect of a dog sticking his head out the window and having his face smooshed by the wind.

Einstein's great breakthrough really came from realizing that the effects of moving through space at high speeds weren't some sort of subtle mechanical effect of the aether but a real distortion of space and time. He showed that the laws of electromagnetism—then the only nongravitational force known—remained completely unchanged if you used the Lorentz transformations. This was absolutely incredible. In one stroke, Einstein not only showed that time and space change with respect to one another according to different observers but also devised a general symmetry principle for all physics to follow. Besides electromagnetism, both the strong and weak nuclear interactions are also Lorentz Invariant (Lorentz almost always gets sole credit), which is just a fancy way of saying that they work the same in any inertial frame. As Einstein described it later in his life:

> All natural laws must be so conditioned that they are covariant with respect to Lorentz transformation.

Lorentz and FitzGerald derived the equations; it was up to Einstein to identify the underlying symmetry of the universe and ultimately explain what those equations meant. Indeed, knowing that they had to be Lorentz Invariant made discovering them much easier.

WHY $E = MC^2$

Relativity fundamentally arises from a symmetry, albeit a fairly roundabout one that involves the relative motion of observers. In the process, time and space are not as independent or as concrete as we might have supposed. But if I were to have asked you ahead of time what the most important part of relativity is, you probably would not have gone for slowed clocks and stretched rulers. Instead, you might be thinking: What does any of this have to do with $E = mc^2$?

Emmy Noether taught us some very important lessons about the nature of symmetry and conservation laws. She showed us, for instance, that time is intimately related to energy and that space is intimately related to momentum. You might even be inclined to think of this as an important SAT analogy-type relation:

Momentum:Energy as Space:Time.

We've just seen that the flow of space and time are a matter of your perspective. Moving observers will measure them differently from the way someone at rest will. But here's the thing about time. No matter what frame you're looking at, time still needs to pass. You can't, just by moving fast enough, cause the flow of time to stop entirely.

On the other hand, because Superman can fly "faster than a speeding bullet," if he exactly matches speed, it could seem to him as though a bullet weren't moving through space at all. Superman, by simply shifting his perspective, can make the momentum disappear entirely.

And now we get to the crux of it. Energy is like time, not space. No matter your perspective, you can't make the flow of my clock stop entirely, and likewise, no matter how you look at a particle, you can't make the energy disappear entirely. We're obviously not going to do a derivation here, but you already have a hint of it. No matter how much you slow down a particle, it'll still have a whopping big energy storehouse of $E = mc^2$.

I've pretended to "derive" Einstein's famous equation in just a few sentences, and you may think I'm pulling a fast one, but the step from discovering relativity in the flow of space and time to the equivalence of mass and energy is a surprisingly short one. Einstein wrote his paper explaining relativity in June 1905 and sent in his follow-up paper deriving $E = mc^2$ in September of the same year.

Einstein's approach was a bit different from the hand waving I've done here, but it was nearly as quick, taking only three pages. He imagined an atom sitting at rest that suddenly gives off two bundles of light of equal intensity in either direction. Einstein himself had discovered the photon only in March, so he refrained from using it in his derivation.

Photons and Relativity

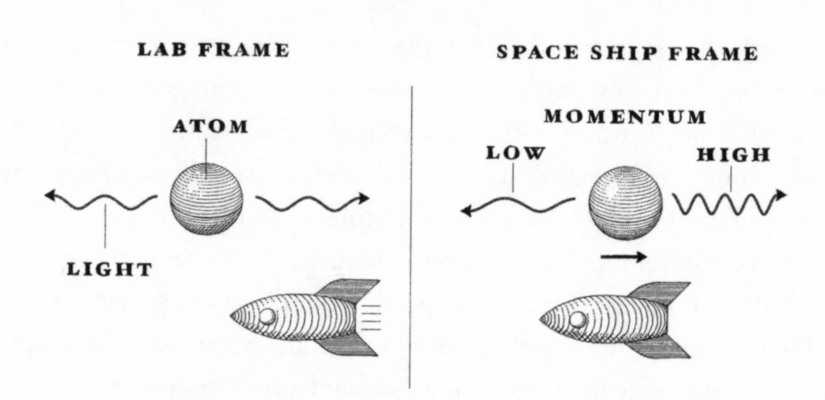

Light is, essentially, pure energy and momentum, and the two are related in a very simple way, by c. This much had been known for 50 years prior. From the perspective of the atom, an equal amount of momentum is carried off in each direction, so the velocity of the atom remains unchanged. Newton and Galileo gave us the conservation of momentum, which means that if the atom started at rest, it would end up at rest after the emission.

There's nothing terribly controversial so far, but Einstein upped the stakes by imagining what everything would look like were you to fly past this experiment. If you run headfirst toward an incoming photon, it won't appear to go any faster than the speed of light—we've already established that. Instead, it *will* appear somewhat higher energy (bluer) than in the stationary frame. Likewise, the other photon will have lower energy (redder). This is known as the Doppler shift, and had been known since the mid-nineteenth century.

But here's the problem. As Einstein did the calculation, he realized that the numbers didn't add up. Momentum is supposed to be conserved, but according to a moving observer, the forward-going photon seems to carry more momentum than the backward-going one. Where did the extra momentum come from? The only possible culprit is the atom, and the only way that the atom could lose momentum without slowing down is if it somehow lost mass—exactly by the amount found from $E = mc^2$.

Bam!

WHY YOU CAN'T HAVE AN ANSIBLE

Remember Ursula K. Le Guin's ansible—the device that could communicate instantaneously across interstellar distances? Similar devices have shown up in everything from *Ender's Game* to the His Dark Materials trilogy. Ansibles also show up in a fair amount of speculative pseudoscience writing. People have tried to imagine jury-rigged entangled particles that instantaneously transmit signals—all without success. As we'll see, quantum mechanics enforces the light-speed limit.

We'd asked at the beginning of the chapter why these devices are impossible, and I'm now willing and able to provide a better answer than "Because I said so."

The magic of relativity is that space and time are intermingled with one another. This means that not only isn't time absolute, there isn't even any way we can say that two things happen "at the same time." To understand why, let's pretend I have a working ansible in my garage. I call my pal Solkar on Vulcan, and he receives my transmission instantaneously. The distance from cause to effect is 16 light-years in space, and zero in time.

We've already seen that distances change according to moving observers, but that the *combination* of space and time needs to be the same for everybody. This means that the time between transmission and receipt *can't* be objectively zero.

For instance, to a ship heading from Vulcan to earth, Solkar will appear to get my message after I send it, exactly as you might expect. To put a few numbers on this, if the ship is flying at half the speed of light, there will be a 9-year lag in the signal. That's not instantaneous at all!

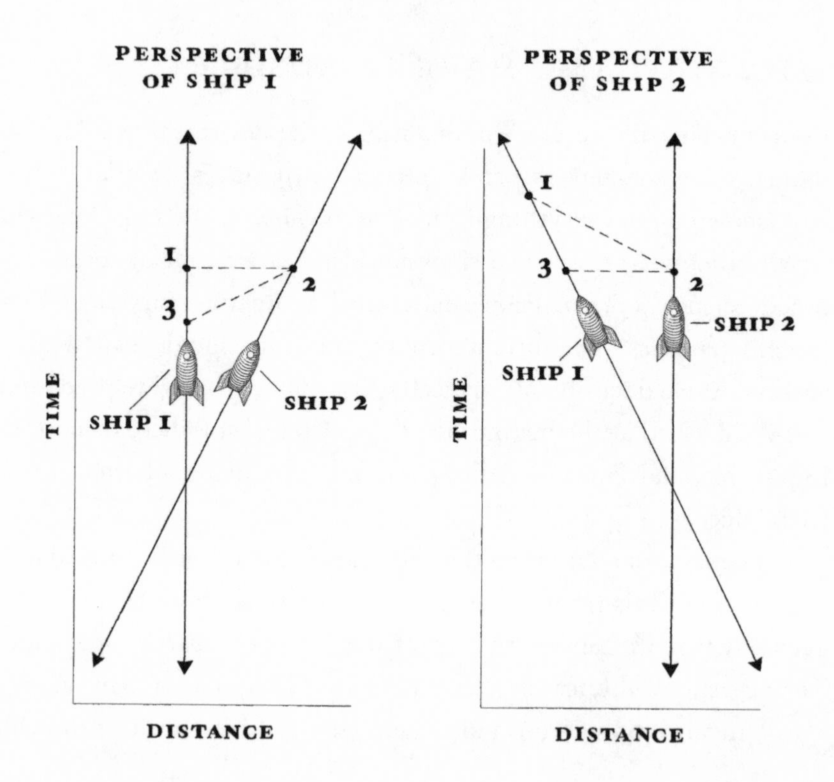

It gets even stranger and more disturbing, though. A ship heading from earth to Vulcan sees the exact opposite sequence of events. Remember, this isn't just an optical illusion or anything of the sort. The earth-to-Vulcan pilot's perspective is completely valid, and yet from his perspective Solkar receives my message a full 9 years *before* I even sent it. If anything like causality is to exist in the universe, this pretty much rules out ansibles.*

Because of the quirks of how velocities add in relativity, you don't even need instantaneous communication to send a message to the past. Einstein himself realized in 1907 that any signal propagating faster than light—tachyons, they're called—could be turned into an antitelephone that could make calls to the past. All you need to do is get in a spaceship and fly away from a friend whose ship is also equipped with an antitelephone transmitter/receiver. Send a message at faster than light and have your friend respond. You will receive your response before you sent your original message. This is just about the most messed-up thing I can think of, though it doesn't prevent legitimate physicists from trying to come up with ansible proofs-of-concept. As a word of caution, if you think you've come up with a working ansible design, please don't call Stockholm; they'll call you.

As a bit of recent scientific history, consider the impact of an announcement made in late 2011, suggesting the (ultimately disproved) possibility of sending signals faster than light. The OPERA experiment in Gran Sasso, Italy, reported a measurement of neutrinos traveling a tiny, tiny bit faster than light. The neutrinos seemed to be going only about 0.002 percent faster than light, but even

* Of course, causality *doesn't* necessarily hold in our universe. While Special Relativity rules out faster-than-light communication, *General* Relativity doesn't. There are a number of speculative designs in General Relativity for objects like wormholes, objects that could be used to violate causality and potentially travel through time. Also, they'd make damn fine ansibles.

that tiny difference would, in principle, be enough to build a giant antitelephone.

The scientific community and the popular science blogs exploded with excitement. Rather than just accept the fact that a century's worth of relativity was unlikely to be undone so easily, a few scientists used the opportunity to propose some fairly fantastical explanations for the discrepancy. Perhaps, went one argument, neutrinos were simply taking a shortcut through additional dimensions of spacetime.

Others (including, I'm proud to report, me) were far more skeptical, and within weeks, literally dozens of papers were sent out to journals suggesting everything from general relativistic satellite effects to statistical anomalies. Something seemed strange from the outset. After all, we had observed astronomical neutrinos for many years. Detectors on earth had measured neutrinos from Supernova 1987 (a whopping 160,000 light-years from earth) within a few hours of the photons. If the supernova neutrinos were as swift as the ones measured by CERN, they should have beaten the photons by 3.2 *years*.

In the end, the OPERA team themselves identified the problem, and it was something incredibly prosaic: loose cables. If there is a moral to the story, it's that you should be very sure about your claims before making a world-changing scientific announcement. The OPERA team did the best it could in trying to identify all possible problems, but the entire affair resulted in the leaders of the collaboration resigning.

The secondary moral, and really the reason I bring this all up, is that actually breaking the speed of light would be a whopping big deal because it really would change just about everything we understand about how the laws of physics work; c is just *that* important.

And even within those limits, it's worth remembering that we're not even remotely close to *reaching* the speed of light, let alone exceeding it. The top speeds astronauts have *ever* reached is about 25,000 mph, as the *Apollo 10* achieved reentry into the earth's atmosphere.

That sounds like some serious speed until you crunch through the numbers and realize that it corresponds to roughly 0.004 percent the speed of light. My point, I suppose, is not to be greedy and to accept the physical limitations of the universe. The same limit that prevents us from reaching warp speeds also, potentially, prevents us from being invaded.

Now, that's a silver lining.

Chapter 6

GRAVITY

IN WHICH WE LEARN WHY BLACK HOLES DON'T LAST FOREVER

Special Relativity was an amazing breakthrough in how we understood the world to work, but even so, it left a lot of questions unanswered, especially when it came to gravity.

Oh sure, Newton had given us a basic picture of gravity a few hundred years earlier, but in the wake of relativity, Newton's version of gravity came off as kind of naive. For one thing, Newtonian gravity moves at infinite speed. If Galactus, destroyer of worlds, devours an entire star system in another part of the galaxy, then according to Newton, we'd feel the gravitational effects instantaneously.

But Einstein put the kibosh on that with relativity. Gravity, no less than electromagnetism, is bound by this speed limit of light. It's almost as if we took a step back with Special Relativity. Before 1905, we were pretty sure we knew how gravity worked. After 1905? Not a clue.

Gravity and light intersect in another way. There are objects out there in the universe, and don't tell me you never heard of them, called black holes. And if there's one thing you know about black holes, it's that they are so dense that nothing can escape—and here's where italics should indicate a deep, ominous, spooky voice—*not even light.*

Black holes are the ultimate gravitational laboratory. On first blush, they seem to be unstoppable eating machines that will ultimately consume the entire universe. But black holes can also be vulnerable; they will even at some point completely fade away. To understand why, we need another foray into symmetry. We'll start with one of the big unanswered questions from Special Relativity.

THE TWIN PARADOX

Within a few years of developing his special theory, Einstein devised a nice little thought experiment that sounds a lot like a first draft for a science fiction story he was working on:

> If we placed a living organism in a box . . . one could arrange that the organism, after any arbitrary lengthy flight, could be returned to its original spot in a scarcely altered condition, while corresponding organisms which had remained in their original positions had already long since given way to new generations. For the moving organism, the lengthy time of the journey was a mere instant, provided the motion took place with approximately the speed of light.

This, with just a little retelling, is the famous Twin Paradox.

The setup goes as follows: Emily and Bonnie are twins. One day Bonnie steps into a spaceship, the starship *Awesome*, and heads to Vulcan at 90 percent the speed of light. She's sent on a simple diplomatic mission, one that requires only a few moments off ship, after which she'll turn around and come home. I know, it seems like kind of a waste to me too.

Twin Paradox

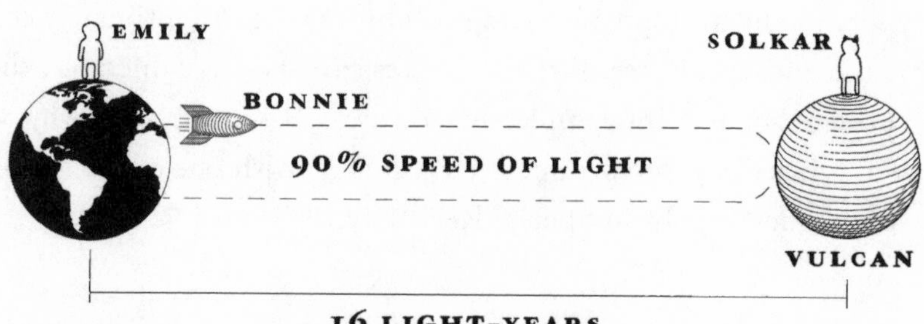

Bonnie completes the 16-light-year trip to Vulcan in a little less than 18 years. That, plus the trip back, means that Bonnie makes her triumphant return to earth in approximately 35 years, 6 months, and 18 days. When she steps out of her ship, she finds that her family and friends have aged accordingly. Surprisingly—in what will be the first of several such reversals—Bonnie appears much younger than her twin. Bonnie has aged only a bit over 15 years and experienced only 15 years of memories. Her shipboard computer has recorded only 15 years of data, and so on. Time inside the *Awesome* seems to have run at a little less than half the rate here on earth, and the distance on the odometer measured less than half of what was initially expected.

While the stretching of time and space are weird, it might not be immediately obvious why this is a Twin *Paradox* rather than simply a Twin "Weird Thing That Happens, but Lots of Weird Things Happen, So Don't Get Bent Out of Shape About It." It didn't, for instance, especially trouble Einstein; he just thought of it as a cool side effect of living in a relativistic universe.

But here's where I bring in a *second* reversal. The first postulate of Special Relativity—the assumption that allowed us to find all of these weird results in the first place—was that we're not supposed to be able to tell who is moving and who is standing still. The perspectives of two astronauts passing each other are perfectly symmetric. Each one feels as though he is stationary and that the other guy is moving. There's no objective way to distinguish one perspective from the other.

Yet it's pretty clear that Bonnie was moving in an objective sense that Emily wasn't. Upon her return, she's younger than her twin. The resolution to the paradox can be found in the fine print. Einstein's postulates were based on the assumption of inertial observers. Bonnie's world isn't described by Special Relativity because she had to speed up and slow down, and Emily didn't.

The physics of speeding up and slowing down—the physics of *acceleration*, in other words—seemed somehow related to the big hole that Special Relativity opened up: how gravity works.

Einstein developed General Relativity to close that gap, to explain how accelerations work. As with the special theory, and just about everything else, General Relativity was built around symmetries. The symmetries are so important in this case that they were the main impetus for Emmy Noether to go to Göttingen and develop her eponymous theorem in the first place.

To understand where General Relativity comes from and how it fits into our overall story of symmetry, I want to spend a few minutes talking about one of the things that bothers the hell out of me in most sci-fi movies.

ARTIFICIAL GRAVITY

Say you want to set a movie in space. There's no gravity in deep space—obviously—which is one of the things that makes it so cool.* It's understandable for a director to want to save a few bucks by envisioning

* You can eat literally all the astronaut ice cream and drink all the Tang you like and never gain a pound.

some sort of "gravity drive," allowing his cast to walk around the set rather than commissioning a real spacecraft.

What bothers me isn't so much the need to include artificial gravity, but how it's done. Artificial gravity might seem to be complicated, or even impossible, at the level of universal translators or warp drives, but it's so simple that Newton could have—and did—come up with a workable mechanism more than 300 years ago. Arthur C. Clarke got it right with *2001: A Space Odyssey*. *Babylon 5* got it right, and it was broadcast on TNT, for goodness sake.

All you need is a spinning space station, and the outside becomes the floor. If you've ever been to an amusement park and ridden on the Gravitron ride, you'll know what I'm talking about. The faster it spins, the stronger the artificial gravity.

The natural state of motion for a particle is to move in a straight

line and at a constant speed. By spinning the ship, supposing the radius of the ship is sufficiently large, you'll feel a comfortable artificial gravity because the floor under your feet is constantly changing direction as you turn. The only design flaw is that if the ship is too small, your head will experience a noticeably lower "gravity" than your feet.

You don't even need to rotate your ship; any acceleration will do the trick. An up-going elevator produces a small artificial gravity as it starts up and a small antigravity as you reach your destination. You could design a very nice gravity drive by simply accelerating your spaceship toward the destination for half the trip and decelerating for the second half. In the beginning, the back of the ship is "down." Your only problem will be the middle of your trip, when the ship starts decelerating and you and all of your stuff are thrown to the front as it becomes the new "down."

I'm not bringing up artificial gravity as an excuse to complain about all of the inaccuracies and implausibilities in movies. I'm not *that guy*, I promise. I bring it up to point out how similar artificial and real gravity are to one another. And just as we can easily simulate gravity, we can do the reverse and simulate weightlessness. In other words, there's a definite symmetry between artificial gravity and the real kind.

Astronauts, as you may know, train for space by taking trips aboard what NASA calls the *Weightless Wonder*, but everyone else refers to as the *Vomit Comet*. An aircraft climbs to high altitude at high speed and then, for want of a better phrase, cuts the engines.* For a short while, the entire aircraft is in free fall, which is just a fancy way of saying that the ship and its contents are subject to the pure, unbridled force of gravity of the earth. Throughout their time in free fall, the astronauts feel weightless. Why shouldn't they? They're falling at the

* I'm glossing over a bit. Because the aircraft still has air resistance, simply cutting the gas wouldn't be enough to put the plane into free fall. But for our purposes, this is close enough.

same rate as the ship, so from a relative perspective, they just float around inside.

Or consider the *International Space Station* (ISS), currently in orbit about 400 kilometers above the earth's surface. We think of the ISS as being in space, but in a sense it's very much in earth's grasp. Gravity 400 kilometers above the surface of the earth is only about 11 percent weaker than on the ground. And yet, if you've ever seen footage, the astronauts float around as if they are weightless. It almost sounds like there is some sort of fraud going on, kind of like when they faked the moon landing.*

It's no trick. Just like the Vomit Comet, the space station is in free fall. In the case of the ISS, the free fall actually takes the form of a nearly circular orbit, but that's still falling as far as gravity is concerned. The only reason that the ISS needs to be in space is to avoid colliding against mountains and so that air resistance won't cause it to crash down to earth. True weightlessness has nothing to do with it.

THE EQUIVALENCE PRINCIPLE

Sir Isaac Newton realized that there was a fundamental symmetry between gravity and acceleration—a relationship that is far more puzzling than it might seem at first glance and one that hints that gravity probably occupies some special place among the laws of physics. Mass, he realized, means two very different things, depending on the context:

> Gravitational forces are proportional to the mass of the body being acted on. Your bathroom scale resists the gravitational force between you and the earth and reads a larger number the more massive you are.

---

* They didn't. I shouldn't have to put in this disclaimer, but you'd be surprised how literally some people take things. Check out YouTube for more evidence.

Mass also means something else, something that has nothing at all to do with gravity. It is also the measure of how hard it is to accelerate something or to decelerate it once it's in motion.

If you don't think this is a big deal, you are dead wrong. Despite the fact that the two are so intimately related, there is no obvious reason that the *gravitational* mass of a body and its *inertial* mass should have anything to do with one another.

And yet, they do. Galileo* is so famous because, among much else, he showed that gravity accelerates objects independent of their mass, by comparing the rate that wheels of different sizes and density rolled down a hill. Newton followed suit by showing that the mass of a pendulum was completely irrelevant to its period, and that the only thing that mattered was the length of the pendulum rod.

To realize how strange the relation between gravity and mass is, let's think about the force of electricity. Electrons and protons have the exact opposite charge as one another, which means that a pair of electrons held a meter apart will repel each other with the same force as a pair of protons held at the same separation.

But the *acceleration* of the particles is another matter entirely. Protons, remember, are approximately 2,000 times the mass of the electrons, which means that they are 2,000 times harder to move. Two electrons held a meter apart will accelerate outward at roughly 26 times earth normal gravity. The protons, on the other hand, will barely budge.

To put it another way, if we were robots held on to our world by

* The story, most certainly apocryphal, is that Galileo dropped objects of different masses from the Leaning Tower of Pisa to compare their rate of descent and found them equal. Air resistance would have almost certainly made the less dense ones fall slower. But you could do this experiment in a vacuum or on the moon and see it with your own eyes. In 1971, the astronaut David Scott dropped a feather and a hammer on the moon and observed them to fall with the same acceleration.

electric fields, it is *not* true that all of us robots would fall at the same rate. The ones with the highest charge-to-mass ratio would fall the fastest.

Gravity and mass have a very intimate relationship. It's one of the many reasons that gravity is utterly distinct from the other fundamental forces. But up until Einstein came along, the relation between mass and gravity was more a curiosity than anything else. No one had any idea of why it should be so.

Einstein devised what he referred to as the Equivalence Principle as a starting point. This is the central symmetry of General Relativity and one that will concern us for most of this chapter. Einstein developed relativity over the course of a decade and described his Equivalence Principle in a great many ways. In the end, he described a weak version and a far more startling version that has come to be known as the Einstein Equivalence Principle (EEP). Roughly speaking, the weak versions say:

> **Weak Equivalence Principle:** Particles moving in free fall are locally indistinguishable from inertial systems.

This is such an uncontroversial statement that Galileo and Newton wouldn't have had a problem with it. It simply says that so long as the ISS is in free fall—even if it's *really* in a gravitational field—it will seem as if there is no gravity inside. A physicist on board could do all sorts of experiments and find all the same results as if he were in deep space.

Or very nearly. The gravitational pull of the earth gets weaker the farther you get from the earth, which means that the earth side of the space station feels a slightly higher gravity than the space side. As a result, there's a very, very tiny tidal effect—about a pound in total over the entire 450-ton station—that acts to "stretch" the station.

Einstein realized that freely falling space stations are telling us something fundamental about how real gravity works. Even by 1907, just 2 years after developing Special Relativity, he'd hit on an even stronger statement of the Equivalence Principle:

> [We] assume the complete physical equivalence of a gravitational field and a corresponding acceleration of the reference system.

In case you missed it, this is Einstein taking the Equivalence Principle to an insane, but ultimately correct, extreme. He argued that there is no measurable distinction between real gravity and acceleration, at least locally. The EEP predicts a lot about how the universe works.

For one thing, if gravity is experimentally equivalent to acceleration, then the strength of gravity can't change over the age or breadth of the universe. If it did, then the ratio of inertial and gravitational mass wouldn't be a constant. But it goes even further than that. It means that in all freely falling or deep-space environments, all experiments should work exactly the same. If Einstein is correct—and remember, this is a *postulate*—physics will behave the same over all of space and all of time.

We saw fairly strong evidence for the Equivalence Principle at the Oklo prehistoric nuclear reactor site in Gabon and with the apparent constancy of the fine structure constant (FSC). The EEP *predicts* the spatial and temporal symmetries that Noether got so excited over. In a very real sense, the Equivalence Principle isn't just a symmetry, it's a sort of *meta-symmetry* that tells us what many of the real symmetries in the universe should look like.

I'll finish this thought with a confession. There must be some sense in which General Relativity (and by extension, quite possibly the Equivalence Principle itself) must be wrong, or at the very least, incomplete. In the most extreme high-energy environments, like at the centers of black holes or at the moment of the Big Bang, relativity and quantum mechanics tell us very different things about how the universe works.

We don't even need to dive into a black hole to see the problem. The famous "double-slit experiment" in quantum mechanics involves shooting a beam of electrons through a screen with two small slits etched out. Because of quantum uncertainty, there is no way to figure out which slit a particular electron travels through: An electron literally travels through both slits at once. This, in and of itself, is kind of nuts, but in the context of gravity, it gets even stranger. If the electron goes through one slit, it presumably creates a very slightly different gravitational field than if it goes through the other.

Someday, and that day is not today, we may have a theory of quantum gravity that tells us whether and how, exactly, relativity breaks down and how to fix it, but for the moment, we simply have to do the experiments. And the experiments seem to tell us that the Equivalence Principle holds.

LIFE IN ANTWORLD

Even without working out any of the details of General Relativity, a process that took Einstein almost an entire additional decade post–Special Relativity, he quickly had an inkling of what the final theory should look like. By using the Equivalence Principle, Einstein came up with a scenario for relating artificial to real gravity, one that I'm going to shamelessly steal.

Imagine life on top of a large spinning disk. This is a lot like the two-dimensional universes that we saw back in Chapter 3—you know, the ones we found couldn't actually support life. Humor me, it's way easier to think about 2-D universes than 3-D ones.

In this universe, there are a bunch of superintelligent ants slowly crawling around on the surface. The queen, Marie *Ant*oinette (sorry), sits perfectly still at the center of Antworld. Her royal court surrounds her in close proximity. To an outsider (you), her courtiers slowly rotate about the queen. They don't know any of this, of course. They just grip so as to not be thrown outward by the tug of the rotating disk. As far as they're concerned, "out" is "down." This, as you may recall, is centrifugal force, and it's the same thing that generated the artificial gravity onboard the *Discovery One* in *2001: A Space Odyssey*.

Antworld

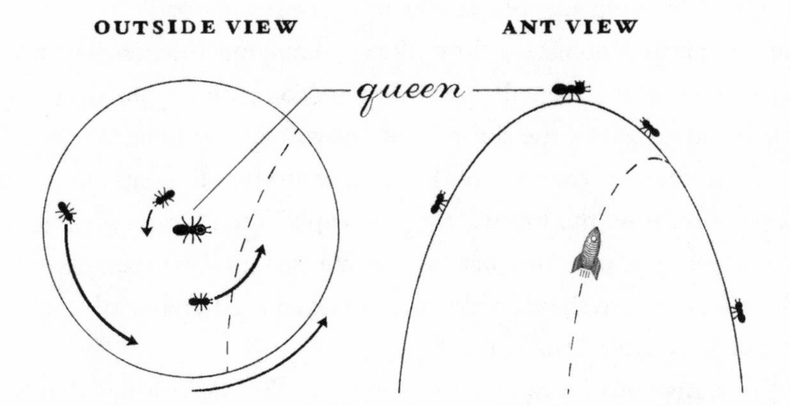

The farther the ants are from Queen Marie, the faster they move and the stronger they are tugged outward. From the perspective of the ants, their Antworld feels very much like a hill with the queen at the top, a hill that gets steeper the farther they go out. An ant that loses its grip will roll outward—down the hill—at an ever-accelerating rate.

There's at least one sense in which this analogy isn't perfect. If you fall down a hill on earth, you'll simply roll down in a radially outward path. An ant falling down the hill in Antworld will start rolling straight down, but will then slowly start rolling *around* the hill as well. This is the famous Coriolis effect. It's the same thing that causes cyclones to spin counterclockwise in the Northern Hemisphere and clockwise in the Southern.*

Because our ants are homebodies, we can ignore the Coriolis effect with impunity. As far as they're concerned, they live on a hill and aren't spinning at all. Outside the Antworld, we know better. The queen isn't

* Despite a common misconception to the contrary, the spinning of the earth does not affect which direction your toilet flows down the drain. Your toilet is just too small.

moving. Nearby ants are moving slowly. Ants farther out move faster. The ants out in the hinterlands are moving fastest of all. This is where all of our training in Special Relativity begins to really pay off. We know something about the flow of time of moving ants. The faster they move, the slower time will appear to pass compared to the queen. The farther out an ant is, the slower that ant will appear to age.

But there's another way of thinking about the same situation, from the perspective of the Equivalence Principle. The *Discovery One* generated artificial gravity by rotating, but if Einstein is correct, except for tidal effects, there shouldn't be any observable difference between rotational gravity and the real thing.

The ants don't know they're moving, so they don't know that Special Relativity should come into play at all. The ants, so far as they can tell, are living in a gravitational field. The ants are very smart. They have discovered something fundamental about how gravity works: The farther you go "down," the slower time runs.

The ant physicists are absolutely correct—about their universe and ours. Time runs slower the closer you get to a massive body, and the more massive the body, the more dramatic the effect. These effects are quite real, but normally ridiculously tiny. Time runs slower on the surface of the earth by less than 1 part in a billion compared to time out in deep space. Over the surface of the earth, the effect is even smaller. Time runs slower at the bottom of Mount Everest than at the top by about 1 part in a *trillion*. Given that we've been confined to the surface of the earth for most of our existence, it's not that surprising that nobody before Einstein noticed that the flow of time changes based on where you are.

But we don't need to travel to other stars to see the significance of time dilation. Global positioning system (GPS) technology in particular requires calibrating satellite clocks and earthbound ones to ridiculous accuracy. Ignoring relativity means that GPS satellites will appear to gain 45 millionths of a second every day. That seems like a tiny number until you note that it corresponds to a positional accuracy of over 10 kilometers.

There are much more extreme environments out there. You could hang out on an incredibly compact, massive stellar remnant known as a neutron star where time runs slower by 20 percent or more. After a decade, 2 extra years will have passed far away. What you've done here is built a (pretty crappy) time machine into the future. But because the gravity on a neutron star is so strong that you'd be squashed like a pancake, traveling to the future is probably the least of your concerns.

There are environments where the relativistic time dilation is even more extreme than on neutron stars. We'll get to the ultimate gravitational environments, black holes, shortly, but first we need to reckon with the fact that time isn't the only thing warped by relativity.

Suppose an outlying ant decides to take a trip around the entire world, a circle. His trip would seem shorter than Queen Marie might have guessed using simple Euclidean geometry. As seen by its residents, Antworld is curved.

What's true for the ants is true for us. As the great general relativist John Archibald Wheeler famously put it:

> Spacetime tells matter how to move; matter tells spacetime how to curve.

Einstein, as you'll recall, couldn't really countenance the idea of a dynamic universe; he mistakenly put his cosmological constant into the equations of General Relativity just to prevent an expanding universe. That wasn't his only error. Much of the development of General Relativity proceeded by a series of steps and missteps.

Hermann Minkowski developed and formalized the concept and mathematics of spacetime in 1908, only a few years after Einstein discovered Special Relativity. Spacetime is a really useful idea, not least because it reminds physicists to always treat space and time on equal footing.

Ignoring either space or time can produce some pretty big errors, especially when dealing with light, which travels through both. For instance, one of the particularly cool predictions of General Relativity is that massive objects will deflect the path of light beams. Stars situated behind the sun will have their light deflected by a small amount, and so Einstein in 1911 predicted the exact amount of the deflection, suggesting:

> As the stars in the parts of the sky near the sun are visible during total eclipses of the sun, this consequence of the theory may be observed. It would be a most desirable thing if astronomers would take up this question.

The next total eclipse wasn't until August of 1914 and would reach totality only in the Crimea region of Russia. Unfortunately for the German expedition sent to observe the events, World War I broke out a few weeks beforehand, and the expedition was captured by Russian soldiers, who confiscated their cameras and equipment. And to add insult to injury, it was apparently cloudy in the Crimea anyway, so even with their equipment, it's unlikely that the eclipse observations would have worked out.

All of this is bad luck for the expedition, but ultimately good fortune for Einstein's reputation. The problem is that in the 1911 calculation, he neglected the "space-like" part of the calculation and was thus off by a factor of two. Had the expedition been successful, relativity would have apparently been discredited, perhaps for a long time. History, as we've seen, is riddled with examples of scientific progress sped up or slowed down by chance meetings, popular perception, or—as in this case—the luck of wars and the weather. The lesson to young scientists is to try to get it right in your *first* published version.

Einstein corrected his analysis in his final synthesis paper in 1915, and during a subsequent eclipse in 1919, Sir Arthur Eddington observed the deflection Einstein predicted. This result cemented in most people's minds the correctness of the theory. It was also a triumph of internationalism following World War I because an English scientist was so instrumental in confirming the theory of a German one.

LIFE NEAR THE EVENT HORIZON

All of the gravitational effects we've seen so far, from Antworld to earth to the gravitational lensing effects of the sun, are small potatoes in the scheme of things. But there are a few places in the universe where the bending of spacetime is overwhelming, such as near the surfaces of black holes.

While the blackness of the black holes precludes us from seeing them directly, we are fairly certain that they are real. Almost every large galaxy, including our own, seems to have a supermassive black hole at the center—some as much as a billion times the mass of the sun—dominating the motions of the central stars.

Black holes are ridiculously simple objects, or at least the nonrotating ones are, which are the only ones I'm going to talk about here. They consist of an infinitely compact "singularity" at the center and an outer boundary known as an event horizon—the point of no return. They are tiny, at least on an astronomical scale. If our sun were to collapse into a black hole, it would be smaller in radius than the city of Philadelphia.* Even the 4 million solar mass black hole at the center of the Milky Way could comfortably fit inside the orbit of Mercury.

* You may be comforted to know that the sun is too small to ever collapse into a black hole of its own accord. Only stars about ten times as massive as the sun will end their days as a black hole. I'm just giving you the numbers for comparative purposes. That said, it's not clear that it's any more comforting for the sun to live

Black holes are the bears of the cosmos. They're dangerous, but they won't hurt you so long as you don't get too close. If the sun were to turn into a black hole, the earth would *not* get sucked inside. Instead, about 8 minutes and 19 seconds after the transformation—the amount of time for light to reach us—you'd see the sun appear to blink out of existence and you'd subsequently freeze to death. But in your dying hours, you'd no doubt be struck by the fact that J. J. Abrams lied to you. Rather than get pulled into the black hole sun, the earth would just keep orbiting that seemingly empty point in the sky, exactly as it always had. Only icier.

But close in, where gravity is strong, it's a whole different story. Mercury is the innermost planet, which means that it feels the sun's gravity more strongly than the rest of us. Even though Mercury is only twice as close in as earth, there are still some hints of how Newton was wrong and Einstein was right.

Newton, and Kepler, found that all of the planets are supposed to orbit the sun in perfectly repeating ellipses. But something seems to be wrong with Mercury, which precesses by about 2 degrees per century. This simply means that rather than make a perfect ellipse, the orbit makes a rosette pattern. For the most part, this can be totally explained by the ordinary Newtonian influences of the other planets, especially Jupiter. But a small effect—about 43 arcseconds per century—is totally inexplicable in Newton's theory. The precession of Mercury makes sense only if you realize that spacetime near the sun is curved.

If the sun were a black hole, we could get much, much closer, and the effects of gravity would be much, much more dramatic. As we've seen, time runs slower near massive bodies than far away. At the event horizon of a black hole, the time dilation becomes literally infinite. Yes, I said, "infinite."

Suppose you had a friend whom you didn't mind sacrificing for

---

out its days as a red giant and ultimately a neutron star than as a black hole. The earth will be toast in any event.

SPAGHETTIFICATION

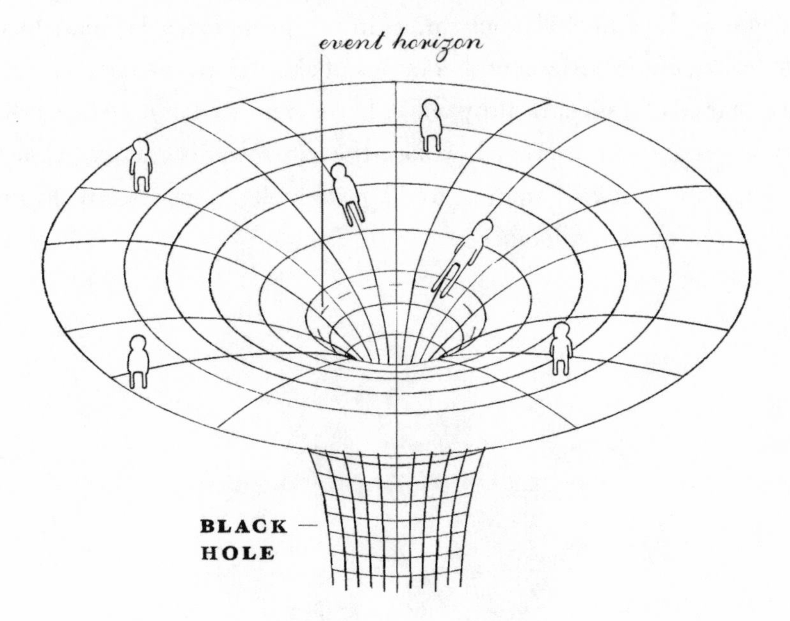

science. Because she's apparently not averse to jumping into rabbit holes and such, let's send Alice. Take a very sturdy rope and dangle her just outside the event horizon. She will think just a few minutes have passed, but by the time she climbs out, thousands of years may have passed for the rest of the universe. By that time, damn dirty apes could rule the earth!

Instead of gently lowering her down, suppose Alice decided to make the ultimate sacrifice and jump feetfirst into a black hole. Just to be clear, I wouldn't suggest that you or any of your friends get close enough to a black hole to conduct this experiment at all. You'd be torn to shreds by tidal forces before you learn anything interesting about the flow of time. This is known as spaghettification, and yes, that's the technical term.

As Alice gets closer and closer to the black hole, the gravity on her feet is going to be a wee bit stronger than on her head. At first, it's no big deal, but as she gets closer and closer, the spaghettification will get stronger and stronger. To put things in perspective, a solar mass black hole has an event horizon with a radius of about 3 kilometers, basically the distance of a nice, healthy walk. If we were to dangle Alice 1,400 kilometers above the event horizon, the *difference* in gravity between her head and feet will amount to 20 *g* of acceleration, already beyond the limit of human tolerance.

As it turns out, more or less regardless of the mass of the black hole, Alice will have about a fifth of a second between the moment when she first notices severe discomfort and the moment her bones are ripped into pieces.

All of this pain can be avoided provided the black hole is big enough. A 10,000 solar mass black hole is far less dangerous than its less massive brethren. The tidal forces are weak enough so that Alice could make it through the event horizon alive. Regardless of how she gets there, once Alice falls below the event horizon, the black hole gains an Alice worth of mass. That's how they grow.

I should probably make a few of caveats about black hole growth. The first is that black holes aren't typically anthropophages. In the wild, their natural diet consists more of dust, gas, stars, and, depending on their size, smaller black holes.

The second is that gravity doesn't arise *solely* from mass. There's an equivalence between mass and energy, remember, so *any* form of energy will ultimately give rise to gravity. I could shine a flashlight into a black hole, and it would, indeed, get a tiny bit more massive. On the flip side, as the sun gives off radiation, it very slowly and very slightly *loses* mass.* The point is that mass is the most efficient way to deliver energy to your system. The right side of $E = mc^2$ is a whopping number even for a small bit of mass, so it's not surprising that the mass contribution to gravity usually gets the most attention.

My third caveat is that *when* Alice falls into the black hole depends a great deal on your perspective. Time near the black hole seems to run slowly—infinitely slowly as you get very close. It turns out that this makes a black hole jump a very oddly one-sided proposition.

As Alice approaches the event horizon, things start getting strange. Even if she survives the forces of spaghettification, we never actually see her cross the event horizon. For one thing, from the outside perspective, it seems to take an infinite amount of time for her

* The sun loses a lot *more* mass from blowing out a continuous stream of particles known as the solar wind, but a tiny fraction of the lost mass is really lost energy.

to actually cross the horizon. Alice doesn't so much cross the event horizon as disappear *Back to the Future* photograph–style, starting with her feet.

If her clock is running slow, this means that everything you could possibly use to measure time, including the frequency of light, will also appear to run slow. Light emitted from Alice's transmitter, for example, becomes longer and longer in wavelength as she approaches the event horizon until you can't see her at all. The effect would be strongest at her feet, and weaker at the tip of her head.

But all of this presents a bit of a puzzle. If time is infinitely slowed near the event horizon, how does anything fall in at all?

The answer, as in so much with relativity, is that it's all a matter of who's making the observation. The gravity we see far away from a black hole is an amalgam of all of the matter and energy that has ever fallen in. The fact that all of that stuff, from our rather limited view, won't ever actually cross the event horizon doesn't really matter. Or, to put it another way, it doesn't really matter how massive the black hole is now, because there's no absolute consensus on what *now* means.

This actually helps us reconcile another mystery that may or may not have occurred to you: If light can't escape a black hole, and gravity travels at the speed of light, how does the gravitational field ever get out of the black hole to let us know that we're supposed to fall in?

Remember Wheeler's explanation of relativity. Mass (and energy) is supposed to tell spacetime how to curve. It's just like a big kid sitting in the middle of the trampoline. It's not the mass of the big kid that makes the little kids roll toward the center, at least not directly; it's the fact that the trampoline is curved toward the center. In grown-up physics terms, as the in-falling material approaches the event horizon, it deforms spacetime as seen from the outside. And *that*, rather than some sort of direct signal, is what causes Alice and everything else to be attracted to black holes.

RADIATION AND PERSPECTIVE

With all of this talk about what goes *into* a black hole, we're going to finish up with a discussion of what comes out. You might reasonably guess, nothing. After all, despite my prevaricating about how gravity works, black holes are called that because light can't escape.

On the other hand, we can "see" black holes in the form of quasars in other galaxies. Quasars are enormously luminous objects at the centers of galaxies where hot, glowing clouds of gas get swallowed up by supermassive black. And by the way, with the exception of giant radio jets, we can't even generally resolve these clouds. When you see detailed accretion disks in news stories about black holes, that's somebody monkeying about with MS Paint or whatever they use these days to make artists' conceptions.

But even ignoring the radiation due to quasars, black holes still aren't *entirely* black. To understand why, let's go back to the *International Space Station*. As we've already seen, the astronauts aboard the ISS believe, quite reasonably, that they are weightless, and everything onboard shares in this fiction.

We could take a Van de Graaff generator and deposit it in the middle of the ISS. This is one of those devices that shows up in old horror movies where a moving belt generates a large electrical charge on a big metal sphere. It's also a surefire sign that you're dealing with a mad scientist.

To the good men and women inside, the Van de Graaff will hang in midair. This is not an especially big deal. The appeal of weightlessness is that stuff generally stays where you left it. It's a simple consequence of the Equivalence Principle.

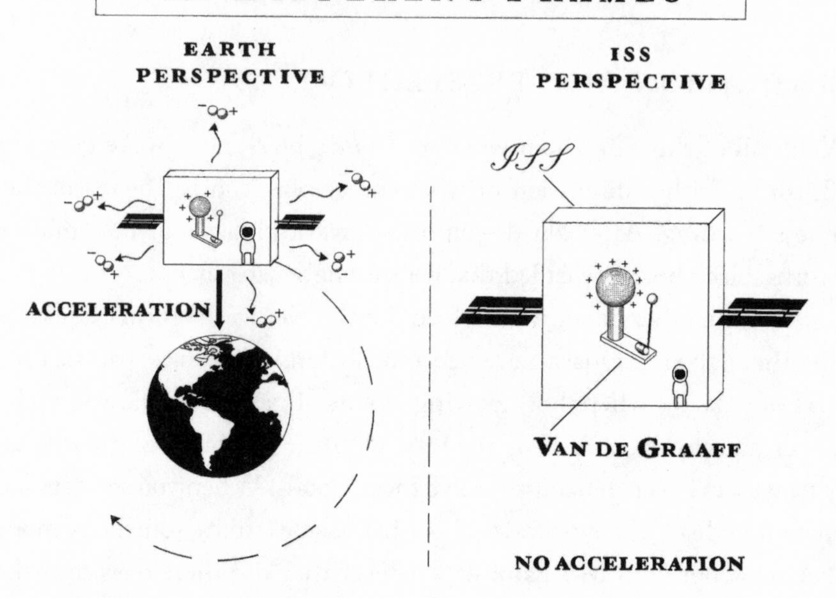

But now think of that same charge as seen from the outside. Let me park my own spaceship some distance away, far enough that the gravity of the earth can be safely ignored.

From my perspective, the astronauts, the space station, and the Van de Graaff speed by at nearly 17,000 miles per hour. But even more important, because the space station, astronauts, and charged generator are constantly changing directions, moving in a circle, in this case, they are clearly being accelerated.

Accelerated charges radiate. A radio transmitter works, for example, by jittering a bunch of electrons at the source and emitting radiation with a particular frequency. Facilities like Brookhaven National Labs on Long Island have giant magnetic rings where they run electrons in a nice circular orbit. The facility at Brookhaven is pretty extreme. With a radius of about 800 meters, and the electrons flying around at about 99.999999 percent the speed of light, the acceleration is *much* stronger than the gentle tug of gravity from the earth. This produces something known as synchrotron radiation and is a very useful experimental light source.

All this is to say that if a Van de Graaff is moving around in a circle—even if that circle is around the entire earth—then it should produce some sort of radiation. But if the Equivalence Principle is correct, then aboard the ISS, every experiment should be consistent with no acceleration at all. The astronauts, by this argument, shouldn't see any radiation coming out of the charge.

This is seriously messed up. Light should either exist or not. It shouldn't be a matter of perspective.*

One could argue that I'm making a big deal out of nothing. After all, one accelerating charge isn't very much, and the gravitational acceleration of the earth is pretty mild, astronomically speaking. But in the real universe, we're *not* just talking about a few charges.

One of the most surprising predictions of quantum mechanics is that even what we normally think of as empty space, the vacuum, isn't completely empty. We are, as it turns out, living in a bubbling ocean of particles and antiparticles. We just tend not to notice them because they are *extremely* ephemeral. To put things in perspective, electron–positron pairs last for only about 10^{-21} second, only enough time for them to travel, at most, a bit more than the nuclear radius of an atom.

Also, because particles and antiparticles always have the exact opposite charge from one another, vacuum fluctuations cancel each other out, electrically speaking.

---

* You may note that I'm leaving this point as a bit of a mystery. That's because it is a mystery. The problem is that *if* the electron really radiates, it will do so with multi-light-year wavelength photons. Not only is that comically huge but it's also much, much larger than both the space station and the scales on which you can safely ignore tidal effects. If you can't ignore tidal effects, then you need to be very, very careful about applying the Equivalence Principle. That said, the general arguments about gravity, acceleration, and radiation that follow will work just fine.

And there's good reason to suppose that this vacuum energy isn't some crazy thing thought up by theoretical physicists just to make the universe even stranger. In 1948, Hendrik Casimir noticed that if you take two neutral metal plates and place them close together, they attract one another. This is known as the Casimir effect, and it basically works only if you imagine that there is a swarm of virtual charged particles between the plates, exactly with the properties predicted by our vacuum energy density.

On the other hand, there *are* reasons to be very wary about the vacuum energy density in the universe. In any given region of space, photons (which are their own antiparticles, remember) of every possible wavelength get created. Because very short wavelength photons also have very high energy, the natural result of all of this is that, at any given instant, there should be an infinite amount of vacuum energy literally everywhere.

For most purposes, we don't *care* about the infinity showing up in our equations. We don't measure energy directly; we measure only difference. It's not a big deal to have an infinity because *everywhere* gets the same infinity. We just subtract it out and hope nobody notices.

We actually can do a little better than that. Physicists work under the assumption that smaller than some scale—the Planck length—the physics that we know will break down. The upshot is that we can make the infinity go away. On the other hand, even if the shortest wavelength photons were at the Planck length, the corresponding energy density would be something like 10^{120} times larger than the actual energy density of the universe. It seems like there is something *very* wrong with this calculation.

Because gravity is supposed to see *all* of the energy in the universe, this factor-of-a-googol error is the worst problem in physics. We're also going to put a pin in it for now. But we *can't* ignore the existence of virtual particle pairs in general because they play a very important role in how black holes work.

Acceleration Through the Vacuum

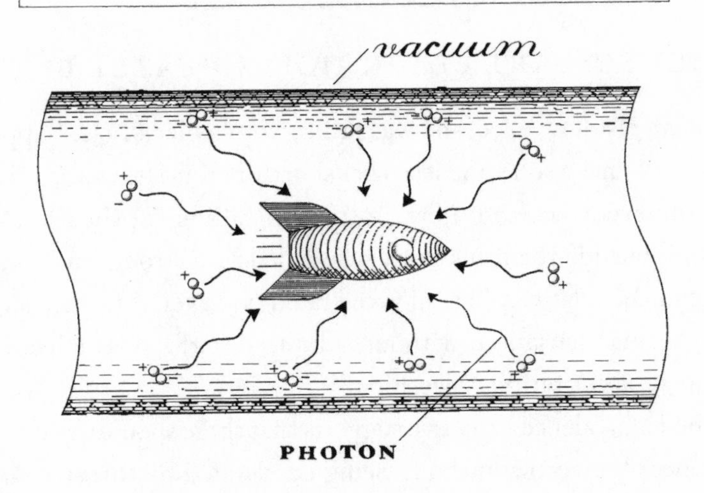

Picture yourself in a rocket ship accelerating through the vacuum, with electrons and positrons constantly created and annihilated. Each of those virtual charged particles appears to accelerate straight at you. And, as you now know, accelerating particles radiate. In other words, if you're in an accelerating rocket ship, the mere act of turning on your thrusters will cause you to see light emerging from the vacuum. If we hadn't done that whole thing with the ISS, you'd probably think I'm full of crap right now.

In the 1970s, several researchers independently found that an accelerated observer will see radiation, including the Canadian physicist William Unruh, for whom the effect is named. Under normal circumstances, this is a tiny effect. Accelerating at *g*, you'd see Unruh radiation of only about 4×10^{-20} K. This is ridiculously cold even by the normal standards of deep space.

I bring up the radiation seen by an accelerating observer, because Einstein gave us a symmetry: There's no real difference between accelerating and being in a real gravitational field. And, as we're about to see, that's going to have an important effect on how black holes really work.

SERIOUSLY, ARE BLACK HOLES *REALLY* BLACK?

When we last left Alice, she was falling into a black hole. Suppose we decide to do her a solid and lasso her shortly before she would otherwise cross the event horizon. Now, instead of falling in, she's just sort of dangling outside the black hole, suspended by a strong cord. She is, as we say in the relativity biz, an accelerated observer. After all, she'd get a very similar sensation, without, admittedly, the tidal effects, if she were in an accelerating spaceship.

The Equivalence Principle suggests that there shouldn't be any local differences between somebody being accelerated by thrusters and one who's really in a gravitational field. Because we'd see Unruh radiation from a rocket ship, Alice should see the same thing hanging around outside a black hole. In other words, she'll see the black hole glow.

In 1974, Stephen Hawking made a foray into combining quantum mechanics and general relativity and showed that black holes are not, in fact, black. This is one of the coolest ideas in astrophysics and one that most physicists believe, even though we've never observed it. All you need to know is that accelerating observers see radiation plus the Equivalence Principle, and *blammo!* You have Hawking radiation.

The Equivalence Principle itself suggests that the laws of physics should be invariant throughout time, and *that*, according to Noether, means we get a conservation of energy. But here's a puzzle: Because radiation is a form of energy, and black holes blast that energy out into the cosmos, the energy has to come from somewhere. There's only one exploitable store of energy in the vicinity of the black hole, and that is, of course, the mass of the black hole itself.

HAWKING RADIATION

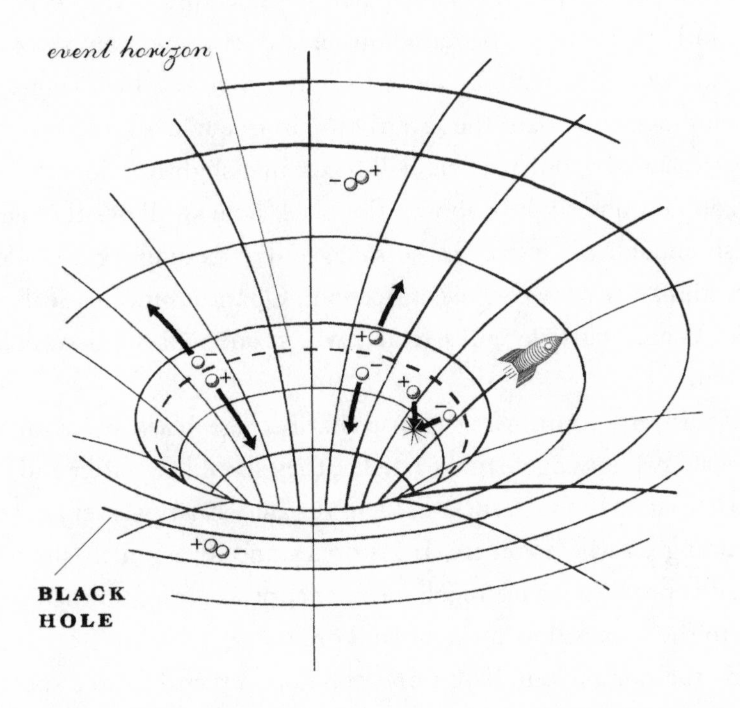

Consider a randomly created particle–antiparticle pair.* Normally, when you create a pair, the two want nothing more than to reunite. And they do so very quickly. One of the big predictions of quantum mechanical uncertainty is that the larger the amount of energy

* Usually when you hear the story of Hawking radiation, people talk about the two particles as an electron and a positron because they have to be antiparticles of one another. Around real black holes, it's actually much more common to create a pair of photons—massless particles of light. Photons are much cheaper to produce than are electrons.

borrowed from the vacuum to make the pair, the shorter they can remain apart. The Force, after all, must remain in balance.

From the perspective of the particles created near the event horizon, nothing seems particularly unusual, at least not from the outset. The particles don't actually *know* that they're near a black hole. Like our astronauts on board the ISS, they're in free fall.

Occasionally, one particle will be created slightly below the event horizon and one slightly above. The black hole swallows the particle foolish enough to violate the cosmic No Trespassing sign, while the other flies away to sweet, sweet freedom. Quantum mechanical fate is fickle. Which particle will live and which one will die is completely random.

Intuitively, you might suppose that because black holes are continuously swallowing virtual particles, they should get fatter and fatter as they feast on the vacuum. But there's a catch. Energy is very different depending on where you are. If I drop a piano from a fifth-story window, it appears to have a much lower energy according to the dropper than to the poor fellow it might land on.

By the same token, if you fire a photon outward from a spot near (but outside) the event horizon of a black hole, it's going to lose a *lot* of energy as it flies farther and farther away. If it were born exactly on the event horizon, the photon would lose all of its energy. This is why—surprise!—light can't escape from a black hole. Someone standing far away really doesn't care how much energy the photon started with, just how much it has when she observes it.

There are two effects that play off one another. High-energy photons are necessarily created very near to each other. At the same time, the closer a photon is to the event horizon, the more energy it loses on its way out. These two effects combine to give a characteristic energy to the photons when they're observed far away. The larger the black hole, the less energy a typical photon ultimately has and the cooler the radiation appears.

But it gets much stranger. A particle created above the event horizon loses most of its energy. A particle created exactly *on* the event horizon loses *all* of its energy and so would contribute nothing to the mass of the black hole. Any particle created below the event horizon actually has a *negative* energy, meaning that when it falls in, the black hole actually loses mass. It's kind of like those "negative-calorie foods," such as celery, that take more energy to eat than the food actually contains in calories.

The amount of mass lost by the black hole as it gobbles the photon below the event horizon is precisely equal to the energy of the photon that escaped (with a c^2 for good measure). Tada! The black hole is well on its way to evaporation.

Let's put some numbers on this so you can impress people at your next cookout or comic book convention or whatever your scene happens to be. A solar mass black hole would radiate at a temperature of about 60 billionths of a degree Kelvin. This is staggeringly cold, about 50 million times cooler than the background temperature of the universe. Because heat flows from hot to cold, the radiation of the universe actually *feeds* a solar mass black hole. Only incredibly puny ones, less massive than the moon, are actually shrinking at present.

Solar mass black holes won't actually start evaporating until the universe gets 50 million times cooler (and thus 50 million times bigger) than it is now. That won't be for a few hundred billion years or so. In other words, there's no chance we could actually use Hawking radiation to see black holes. They're just too cool.

The very fact that black holes have a temperature at all might be a bit surprising. As we've seen, temperature is supposed to be related to entropy, and it's not clear what entropy should even mean when you're talking about an object that swallows everything that it comes into contact with. After all, how many different ways can you arrange an unstoppable eating machine that's defined only by a single number—mass?

What's more, the amount of *heat* that falls in is dramatically different depending on what falls in. Complicated systems like pennies have low entropy, whereas blobs of warm gas have very high entropy, and yet they contribute the same mass to the black hole when they fall in. Where does that information go? As Hawking himself has said:

> If you jump into a black hole, your mass energy will be returned to our universe but in a mangled form which contains the information about what you were like but in a state where it can not be easily recognized. It is like burning an encyclopedia. Information is not lost, if one keeps the smoke and the ashes. But it is difficult to read.

Black holes are essentially entropy-producing machines. The Second Law told us that entropy was supposed to increase in general, but what Hawking predicted was that no matter what you put in, the black holes are literally the most disordered things that physics will allow.

The entropy contained just in the central black hole in our own galaxy is greater than all of the entropy within the observable universe during the Big Bang. And that's only one black hole. The ultimate fate, and the only way in which entropy will be able to increase yet more, is for it to evaporate into a ridiculously huge number of low-energy photons.

A simple symmetry—the Equivalence Principle—gives us Unruh radiation and black hole evaporation. At first, this seems like nothing more than a cute example of some fairly abstract astronomical objects, but in reality, it's a terrifying glimpse into the future of the universe.

Stars, planets, and gas will all spiral and fall into black holes, and those black holes will disappear a few quadrillion years later. Meanwhile, the universe will continue accelerating and accelerating, isolating each cold cloud of gas from others forever. After that, a cold, dead universe filled with nothing but entropy. That's the price you pay for a unidirectional arrow of time. Symmetries govern not only the beginning of the universe but the end as well.

Chapter 7

REPLACEMENT

IN WHICH WE CONSIDER THE DESIGN SPECIFICATIONS FOR A TELEPORTATION DEVICE

I think it's time we had a serious talk. For a while now—too long, maybe—I've let you live in a classical universe, one in which quantum mechanics intruded only long enough to mess things up a bit. But now it's time for hard, cold reality. Quantum mechanics isn't just a minor detail; it dominates the universe. And symmetry at the quantum level is even stranger than at the classical one. At the quantum level, as we'll see, we won't even be able to distinguish between *this* particle and *that* particle. At the risk of sounding too new age-y, it's all one.

Particles are interchangeable, and, sorry, so are you. All of the seemingly enormous differences in you, me, and everybody else are fundamentally attributable to such a small thing as the arrangement and number of the protons, neutrons, and electrons in our respective bodies. We tend to think of our atoms and molecules as being "ours" in some fundamental way, but really, we're just borrowing them. To give you an idea of both how numerous our constituent particles are and how willing we are to share them, take a breath.

A breath seems like a relatively small thing, but with each inhale or exhale, we move around something like a hundred million quadrillion molecules. Consider the dying accusation of Julius Caesar, "Et tu, Brute?" Caesar is a good example because he's been dead long enough that his breath has had the chance to be scattered around the world, and every time you inhale, you typically breathe in about one molecule from Caesar's final words. For that matter, you're currently sharing breath with just about anyone who either lives near you or has been dead more than a few hundred years.

The point is that your molecules aren't really yours, no matter how you'd like to think of them. Every year, about 98 percent of your atoms are replaced with others exactly like them. The comedian Steven Wright once joked:

> I woke up one morning and all of my stuff had been stolen and replaced by exact duplicates.*

It's funny because it's true. Identical atoms more than just guarantee that you can find a replacement carbon if you need one. This is going to form the basis for a fundamental symmetry in the universe:

> **Particle Replacement Symmetry:** All of the measurable quantities in a system will remain unchanged if you swap two particles of identical type and state.

Particle Replacement Symmetry is a surprisingly important, and very real, symmetry of the universe. It provides the basis of teleportation, the ends of stars, and ultimately all of chemistry.

---

* I know. It's mostly in the delivery.

HOW TO BUILD A TELEPORTER

If you've ever seen *Star Trek* (and you have), you're already familiar with the idea of the transporter. You stand on a pad, your atoms are ripped apart and then reconstructed elsewhere. The reconstructed entity has your looks, your memories, and atoms in identical configuration to the original.

If television has taught us anything, it's that any time you entertain the possibility of duplicating people, the copied, goateed versions are going to be evil. But both you and your clone are built identically down to and including our constituent atoms. Your crewmates have to kill the evil version, but how can they tell the difference?*

Star Trek could *almost* be a documentary sent from the future about how teleportation really works. Gene Roddenberry got only a few details wrong. In real teleportation your atoms aren't sent from the transporter pad down to the planet. Instead, following the teleportation, there's a you-size pile of chemicals on the "from" pad, and the destination pad builds a new you out of a chemistry set at the other end. Beyond that, *Star Trek* could totally happen.

Awesome, right?

The philosopher Derek Parfit describes a device very similar to what we've been talking about (you know, without all of the evil):

> The scanner here on Earth will destroy my brain and body, while recording the exact states of all of my cells. It will then transmit this information by radio. Traveling at the speed of light, the message will take three minutes to reach the Replicator on Mars. This will then create, out of new matter, a brain and body exactly like mine. It will be in this body that I shall wake up.

* If you've noticed that I've totally mashed up the plot of Episode 33 of *Star Trek: The Original Series* with half a dozen other parallel universe romps, congratulations, you graduate from nerd academy.

Despite the fact that your body is now made of entirely new atoms, because they are in an identical configuration to the original, you could absolutely argue that the person who wakes up in the replicator really is you.

Parfit ups the stakes by then asking what happens if the original isn't destroyed. Which you is you? As we're going to see, this issue doesn't actually come up in "real"* teleportation devices because the original will always be destroyed. Still, this scenario should give you pause. If you aren't the sum of your particles, then *what* are you?

In 1993, a computer scientist at IBM named Charles H. Bennett proposed the first practical teleportation device, where *practical* in this case is really in the eye of the beholder. You get to transport only a single particle at a time.

I figured that would be a bit of a downer. Under those conditions, we could presumably just save ourselves a lot of effort and simply do the sleight of hand like your uncle pulling a quarter out of your ear or grabbing your nose. Couldn't I just *say* that I've transported an electron across the room and show you some random electron from a balloon that I'd rubbed on my slacks? How would you know the difference?

UNCERTAINTY AND SPIN

One electron is just as good as another, but that doesn't mean that any two always *appear* the same. As we've seen already, electrons, and all other particles, have an intrinsic property called spin. The simplest teleportation device might simply be a matter of determining which way an electron is spinning, and copying that information over to another electron across the room. Sounds easy, right?

Wrong.

* I put in the scare quotes only so as not to offend your delicate sensibilities. Teleportation happens, just not on human scale. At least not yet.

We're playing in the big leagues now, so it's time to come clean about the nature of quantum mechanics. So far, we've more or less been able to largely ignore the effects of the quantum mechanical world, but I haven't really said what quantum mechanics *is*. It can be summed up in three simple ideas:

1. Physical measurements can have only certain possible outcomes. It's like flipping a coin: The result is either heads or tails and never something in between.
2. There's a random element to the universe. When we measure the energy or spin or position of an electron, we can't say for certain what we'll measure before we actually do the measurement. We can just describe the probabilities.
3. Probabilities are described by waves. The *mechanics* of quantum mechanics simply details how these probabilities of various outcomes change over space and time.

These rules have some profound implications, including one you've likely heard of. In 1927, the German physicist Werner Heisenberg put forth his famous Uncertainty Principle. Heisenberg argued that the better you know where an electron is, the less you know where the electron is going, and vice versa. Uncertainty also means that merely by trying to figure out what the electron is doing, you're going to change it.

For instance, it's possible to set up an electron with a random spin. We know from Chapter 1 that if we actually measure an electron's spin using a set of magnets, we're able to find only one of two possibilities: up or down. We just can't predict which.

The randomness of spin goes far deeper than the randomness of a flipped coin. With a coin, you *could* figure out whether it's going to come out heads or tails, at least in principle. You might measure the wind speed, and force on the coin, measure the exact balance ahead of time, determine the elastic properties of the table you're flipping it onto and so forth. Take all of that data and model them in a computer, and you could predict the flip every time. We consider a coin flip random only because we're too lazy to do all of that work.

With a randomized spin, even the universe itself doesn't have any idea how things are going to turn out. You can't predict whether an electron is spin-up or spin-down because it *isn't* either of those things. Until you make a measurement of them, an electron will quite literally be a combination—a *superposition* as the physicists like to say—of both. The French physicist P. A. M. Dirac did as much as anyone to uncover the fundamentals of spin. As he put it:

> This statistical interpretation is now universally accepted as the best possible interpretation for quantum mechanics, even though many people are unhappy with it. People had got used to the determinism of the last century, where the present determines the future completely, and they now have to get used to a different situation in which the present only gives one information of a statistical nature about the future. A good many people find this unpleasant. . . . I must say that I also do not like indeterminism. I have to accept it because it is certainly the best that we can do with our present knowledge.

How does this indeterminism play out in practice? We once again ask Alice* to help us out in the name of science. She has an electron that she'd like to teleport to a friend. But remember, the only thing that distinguishes one stationary electron from another is its spin. Suppose

* The standard notation in discussions of teleportation (including in Bennett's paper) and cryptography generally is to call the pair Alice and Bob. Fortunately for us, we know just where to find Alice, dangling and spaghettified outside a black hole.

Alice's (unknown) electron is 80 percent spin-up and 20 percent spin-down. How can Alice figure out the odds?

With a weighted coin, it's simple. Flip it a gazillion times, count the number of times it comes up heads, and there's your probability. But with quantum spin there's an added layer of complexity. Once Alice takes a measurement, she alters the system. In this case, 80 percent of the time, she'll measure the electron to be spin-up. But in measuring the electron, she's changed it. She wasn't in control of *how* she was going to change it, but she changed it anyway.

Alice started with an electron that was 80 percent up and 20 percent down, but if she measures it as up (for instance), she'll convert the electron into 100 percent up. Physicists call this the "collapse of the

MEASURING SPIN

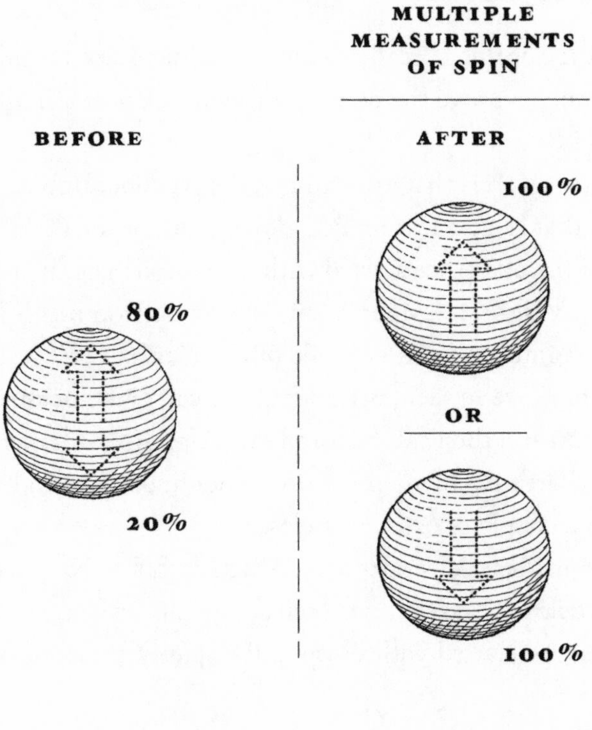

MULTIPLE MEASUREMENTS OF SPIN

BEFORE

80%

20%

AFTER

100%

OR

100%

wavefunction." Because the only possible measurements she can make are up or down, she can't ever get at the mixture in the original electron, which means that she doesn't have enough information to teleport it.

It's a general rule that you can't measure a system without changing it. As Heisenberg himself put it:

> Every experiment destroys some of the knowledge of the system which was obtained by previous experiments.

How do we get that knowledge back? It turns out that a lack of knowledge isn't going to prevent us from teleporting particles. We just need to know how to disentangle some fairly complicated quantum mechanical knots.

ENTANGLEMENT

A practical teleporter is really much more like a fax machine than a particle beam. To prove the point, Alice will teleport a single electron to her friend Bob.

The implication is that after all is said and done, Bob ends up with an electron that is a perfect replica, down to all of the details of quantum spin, of the one Alice started with. If we don't get all of the details of spin exactly right, a teleportation device won't do much more than turn a person into a person-size pile of chemicals.

But spin, as we've seen, is tricky. Alice can't just measure the spin of her electron and then call Bob and say, "Up." Measurement changes everything. Fortunately, there's a back-door approach, and it involves sneaking in a couple of extra particles.

To generate their helper particles, Alice and Bob need to start with an unstable particle with no spin that then decays into an electron and a positron. Because we started with no spin, the spin of the positron must be opposite of that spin of the electron—they have to add up to zero. This is a very simple example of a phenomenon known as entanglement. Measurement of the electron automatically tells you something about the positron.

On the face of it, entanglement sounds trivial. Think of it as the world's most terrible magic trick: I put a black marble and a white marble in a bag. If we each draw out a marble in our closed fists, and I see that mine is black, I *know* that yours is white. Tada!

Einstein famously said, "God doesn't play dice," thinking (wrongly, as it happens) that the spins of electrons are just like black and white marbles. There is no randomness, he claimed, just information that we don't have.

Einstein's gripe about not playing dice wasn't just that quantum mechanics was random but that it seemed to be nonlocal. Special Relativity says that we can't exceed the speed of light without the possibility of messing with causality, but on the face of it, entanglement seems to do an end-run.

On the other hand, if you accept that (1) the spin of Alice's positron is fundamentally random and (2) Bob and Alice will *always* record opposite spins, no matter how far away they are from one another, the only logical conclusion must be that some sort of signal traveled faster than light. To Einstein, faster-than-light communication was a complete nonstarter, so he reasoned there must be *something*, a *hidden variable* as he called it, that preprograms the electron and positron to coordinate so that the spin is always opposite one another. Otherwise, how does the pair *know*?

Einstein's objections remained untested until the 1980s, when the French physicist Alain Aspect and others showed experimentally that there was no possible program, no matter how complex, that could account for the behavior of entangled particles.

This wasn't necessarily the result that John Bell—the physicist who provided the theoretical framework for Aspect's experiments—or anyone else was hoping for:

> For me, it is so reasonable to assume that the photons* in those experiments carry with them programs, which have been correlated in advance, telling them how to behave. This is so rational that I think that when Einstein saw that, and the others refused to see it, *he* was the rational man. The other people, although history has justified them, were burying their heads in the sand.

Something about quantum mechanics allowed *something* to coordinate particles without the light barrier getting in the way. In our teleporter, suppose Alice gets the positron and Bob gets the entangled electron. Half the time, Alice measures the positron as spin-up and Bob measures the electron as spin-down, and half the time, each measures the reverse.

PRACTICAL TELEPORTATION DESIGN

We've already talked about the impossibilities of ansibles, so I don't want to harp on this, but it bears repeating. No matter how clever Alice and Bob may be, they won't be able to get a single bit of information to travel faster than light. Entanglement will not allow you to break that barrier.

If Alice measures spin-up, for instance, she doesn't know whether she made her measurement first and Bob will subsequently measure spin-down, or whether Bob already measured spin-down. And she didn't even know that he'd measure down until she made her own measurement. More to the point, it's not like she had any control over what direction she would measure.

Likewise, Bob can't simply snatch the electron out of the air and turn it to spin-up. Or rather, he can, but that won't send any sort of

* We've been talking about electrons so far, but you can entangle just about whatever you want, including photons.

message to Alice. If you mess with the electron, the pair simply isn't entangled anymore. This is known as decoherence, and it's simply a fancy way of saying that entanglement doesn't last forever. And by *not forever*, I mean that the longest entanglement timescales are tiny fractions of seconds.

Nevertheless, we *can* use entanglement as the central engine for our teleporter. Rather than talk about spin, let's bury all of the quantum details under the hood. Alice and Bob each have a little electronic box. Each box has a button and small screen that can say either "Up" or "Down." Their boxes are connected to one another by a meter that currently reads "Opposite."

The boxes can be programmed with a random number generator, specifying the (unknown) probability of measuring an up. But once Alice presses the button, the original program gets deleted, and a new one, one that always gives the same answer, is written. In the beginning, all we know is that if Alice pushes her button, either "Up" or "Down" will appear on her screen. Subsequently, when (and if) Bob pushes his button, he will get the opposite result from Alice.

How can these devices be used to fax the program in a third box?

This is going to look trivial, but bear with me. Suppose that the third box, call it C, has a really simple program. Unbeknownst to Alice, but beknownst to us, pressing C will give "Up" 100 percent of the time. Just to be clear, C represents the program (particle) that Alice wants to teleport to Bob.

Alice then connects box C to her original box A. Upon connecting the two boxes, the meter on the AC cable reads "Same." The wave function collapses, clearly changing *something* about the program in box C.

This isn't as strange as you might think. Lots of physical systems allow us to make a measurement of the combination of two things without either of them individually. For instance, two magnets held side by side will be at lower energy if they are pointing in opposite directions. Sometimes, you can measure very easily that they are in

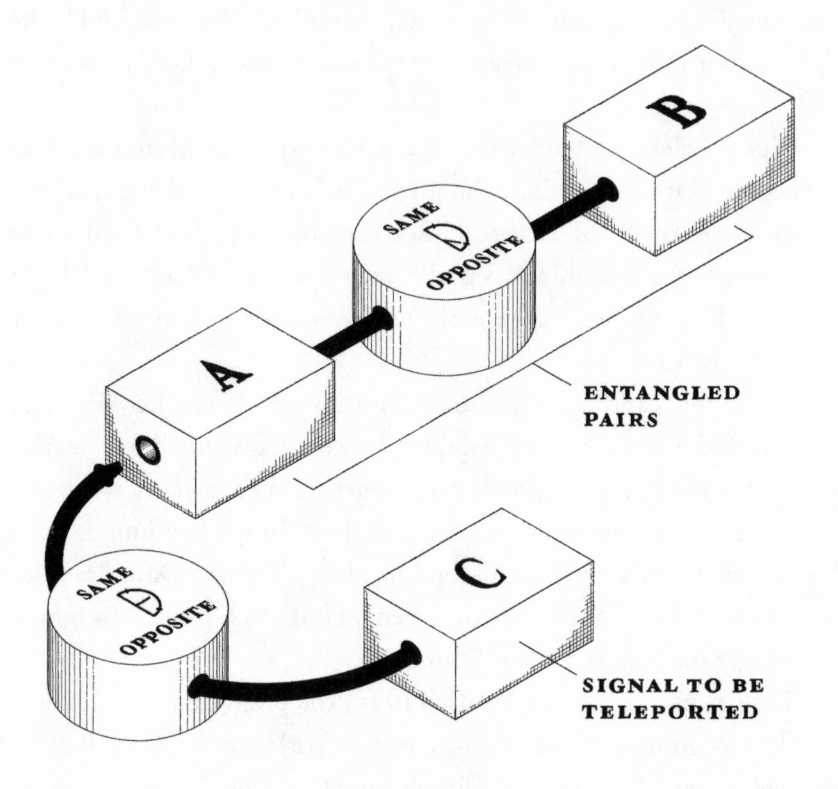

opposite directions, even if you have no idea which direction either of them is pointed in.

We know that B is the opposite of A, which in turn is the same as C. If Bob were to push his button, he would read "Down." It's the only logical possibility.

This entanglement happens instantaneously, which seems like Bob gets his signal faster than light. But wait! Alice now needs to pick up the phone and actually call Bob to tell him, "Boxes A and C are in the same state."

It could have worked the other way. Had the meter read "Opposite"—and, remember, this is a totally random thing—then B is the opposite of A, which, in turn, is the opposite of C. Bob's box is a perfect copy of Alice's original.

In other words, to figure out whether you've sent a positive image or a negative image of the original, you need to measure particle C and then send that info to your friend with ordinary speed-of-light radio signals.

This is straightforward enough if the boxes are in a pure state, but remember that box C could be a combination of Up and Down. The example from earlier gave an 80–20 mix. Alice doesn't know the contents of the box, and if she pushes the C button, she'd select one of the two options randomly and delete the original program. Likewise, on the receiving end, Bob doesn't actually need to measure his copied program. Just knowing that he has an exact duplicate of the original is sufficient.

In practice, teleportation is far more complicated than I'm making it sound.* We've seen that outside stuff tends to interfere with whatever it is you're trying to measure and will cause your signal to decohere. Isolating and measuring the states of individual electrons or atoms is tough enough that the two physicists who mastered it, Serge Haroche and David Wineland, were awarded the 2012 Nobel Prize in physics.

If you're paying careful attention, you may notice a side effect of the teleportation. After all is said and done, you've monkeyed around with the original particle so much that you've destroyed its state. This isn't a result of carelessness. In the early 1980s, several groups showed that no matter how you try to build your quantum teleporter, you will *always* destroy the original. This is known as the No Clone Theorem, and given the prospect of evil twins popping out of teleporters, it's the only thing that allows me to sleep at night.

Even if you're not terribly interested in teleportation, this same reasoning is going to play a pivotal role in emerging technology. While

* And yes, I realize that it doesn't sound simple at all.

our old-school classic computers use bits—1s and 0s—to store information, *quantum* computers will use the magic of entanglement to store and process data in qubits. The qubits will be *superpositions* of 0s and 1s, capable of factoring giant numbers and breaking encryptions incredibly quickly.

And lest you think this is all idle speculation or entirely in the domain of sci-fi, in 1997, an Italian and British group conducted the first successful test of quantum teleportation by teleporting a photon a distance of about 2.5 meters. More recently, in 2012, a Chinese group smashed the previous record by teleporting photons nearly 100 kilometers.

Please restrain your excitement. For one thing, it's only a single particle. For another, even the best quantum teleporters have only an 89 percent fidelity rate. Do you really want to be in the 11 percent that gets destroyed in transit?

Fundamentally, teleporting a car or a person would work exactly the same way. The quantum state of a person, or even a molecule, is just staggeringly more complex than a photon because the amount of information to transmit goes up geometrically. Sending a person is complicated almost beyond reasoning. Almost.

WOULD ANYONE NOTICE?

The real reason that teleportation is so interesting is that it brings to the forefront the question of what it really means for two things to be "the same." This is a book about symmetry, after all.

Electrons *really* are identical, as are protons or neutrons, or any other particle of a particular type. Consider how this plays out in the context of radioactivity. Cosmic rays in our atmosphere continuously make a material known as carbon-14. You may have heard of it. It's pretty famous because it can be used to determine the age of shrouds, old books, and the like. You could imagine Indiana Jones using carbon-14 if he were an actual archaeologist rather than a grave-robbing marauder.

Under most circumstances, carbon-14 behaves just like ordinary carbon. Plants can take it in during their respiration. We eat plants, and it becomes part of us. Mufasa looks on approvingly.

But carbon-14 isn't entirely stable; it decays, on average, after about 6,000 years, breaking down into nitrogen and a few other particles, the details of which needn't concern us overly. When I say "on average" I mean that the decay of carbon-14 is entirely random, and if I had a big lump of the stuff, after about 6,000 years, half of the atoms in that lump would have decayed, and the other would be unchanged. This is, as you may recall, the *definition* of a half-life.

Suppose a particular atom of carbon-14 still hadn't decayed after 5,999 years and we compare it to a brand-new atom just produced in the atmosphere. Which do you suppose would decay first? Your gut reaction might be that the older atom will probably decay sooner, that it's somehow due to decay any day now. But why should it? The magic of Replacement Symmetry says that there's no expiration date marked on the first one to indicate that it's been sitting around awhile.

We could even take this a step further. Supposing I got bored during the millennia that I'm waiting and watching for individual carbon-14 atoms to decay. A mischievous imp (you) comes along while I'm distracted and switches two of the atoms, one newly created, and one nearly 6,000 years old.

For atoms, we've already established that this switch doesn't result in anything obvious. We didn't know which atom would decay first before you did the switch and we still don't. But more important, if you were extremely clever in your swap and made sure that the states of the two atoms were perfectly swapped as well and if particle replacement is an absolute symmetry of the universe (it is), then there is no physical mechanism whatsoever that would tell me that the switch took place.

I mentioned a few rules for quantum mechanics at the top of the chapter, and it's time to return to them. In particular, think back to the quantum wave. If we could describe the wave function of the entire universe, at any given instant, it would give us all probabilities of finding anything anywhere. You could imagine that the universe itself is the *only* supercomputer powerful enough to do the calculation.

But now suppose we swap one atom for another, putting each in the exact same quantum state as the original state of the other. No experiment in the universe could distinguish the swapped version from the original—a perfect symmetry in the sense that Hermann Weyl described it at the beginning of the book.

Replacement Symmetry is one of the most important steps in structuring our understanding of the universe—an ordering that will lead us into the next chapter and ultimately into an explanation of heavy elements and all of chemistry.

To get a running start, we need to know a little about how waves work, whether they are sound waves, light waves, water waves, or quantum mechanical probability waves. The one rule that applies to all of them is that if you double the amplitude of a wave, the strength (the loudness of the sound, or the brightness of the light, or the probability of detection) goes up by a factor of four. Or, to put it more precisely, the strength of a wave is proportional to the *square* of the amplitude.

There are exactly two possible ways to keep the probabilities the same after the switch as before it, either by multiplying the wave function by one or by multiplying by negative one. Some particles take the positive route, and some take the negative route, and which is which will turn out to have some enormous implications.

It's not going too far to say that you owe your very existence to a minus sign. But to understand why, we're going to have to delve deeply into the realm of spin.

Chapter 8

SPIN

IN WHICH WE INVESTIGATE WHY YOU AREN'T A SENTIENT CLOUD OF HELIUM AND WHAT A SPOONFUL OF NEUTRON STAR WOULD DO TO YOU

Picture yourself at a fancy dinner party, and seated next to you, a physicist. This may require some imagination on your part as we don't typically get invited to many of these affairs. To put your newfound physicist friend at ease, after listening to him ramble on for over an hour, you decide to throw him a bone and describe his theory as "elegant." You won't need to say another word. You've just made a friend for life.

While we seldom achieve elegance in our dress or manner, elegance is absolutely at the heart of symmetry and at the heart of physics. And despite the laundry list nature of particle physics, the inhabitants of our particle zoos are remarkably simple creatures. Just a few numbers tell you everything you could possibly want to know. There are the obvious: mass and charge. There are the less obvious: color and flavor, which we'll get into in the next chapter. And then there is the extremely subtle: spin.

Spin is subtle enough that this is the one chapter in which you're not going to get a new symmetry. Sorry. But the good news is that we're going to see how the Replacement Symmetry has very different implications for different particles, and more important, what makes one type of particle behave differently from another. Also there are a bunch of explosions at the end, so stay tuned.

We've seen spin before, but I glossed over some of the quantum insanity. We saw way back in Chapter 1 how the direction of quantum spin, at least for neutrinos, is one of the surest signs that we haven't all been somehow converted into antimatter. The time has come to get to the heart of (the) matter, and to show, once and for all, that you owe your very existence* to how and whether subatomic particles are spinning.

Spin sounds so much like something from your ordinary experience that we might be tempted to think that we understand it already. I can tell you right now, if you're picturing an electron like a microscopic globe or bowling ball, then you're not doing it right.

WHY SPIN IS *NOT* LIKE PLANETARY ROTATION

I'm going to open up my big box o' analogies, and pretend that particles kind of, sort of behave like the spinning earth, mostly because you are (presumably) from the earth and have perhaps owned a globe at some point. And then we'll show how they absolutely *do not*. I just wanted to warn you in advance.

First, silly question, Which way is up? Unless you're standing on the surface of the earth, up is far from obvious.

When ships encounter one another in the movies, they are almost always oriented so that the "up" of the various bridges correspond to the "up" of the screen. How do they know? "Up" in space is pretty much meaningless. That nearly every globe and map of the earth is oriented

* It's really quite amazing how many different aspects of nature you owe your existence to. But spin is a biggie. Without spin, as it turns out, you'd have no more structure than a cloud of helium.

with north as up and south as down is a simple accident of history. The people with the biggest militaries ended up conquering other people, and put their countries on top.* But planets, like particles, have their up defined in another way.

When viewed from the North Pole of the earth (up on globes and maps), the earth rotates counterclockwise. This is just a convention, albeit a very persistent one (I'm told maps in Australia can be had with this convention turned on its head). Likewise, the planets in the solar system have a common history, which means that as they collapsed into nice spheroids, they all ended up orbiting the sun in the same direction. A perfectly useful, albeit arbitrary, definition of up in the solar system

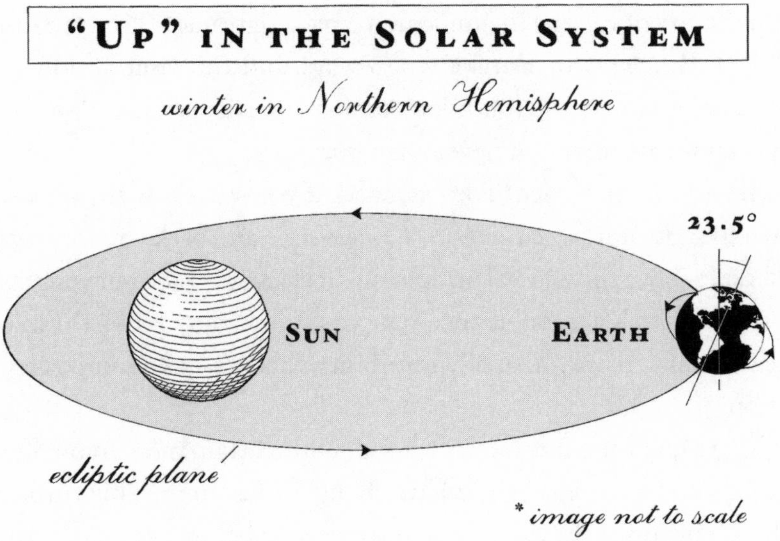

"UP" IN THE SOLAR SYSTEM

winter in Northern Hemisphere

23.5°

SUN

EARTH

ecliptic plane

* *image not to scale*

* That's not the only way that maps can play with your brains. The Mercator map projection significantly magnifies regions nearer to the poles. As a result, Europe and Africa appear roughly the same size, when in reality the latter has three times the land as the former.

can be established by demanding that the planets are all orbiting counterclockwise when seen from "above." With the exception of the utterly disgraced Pluto,* the remaining eight good (or real) planets all orbit the sun in nearly circular orbits in what's known as the ecliptic plane.

The planetary ups and the solar system up don't necessary align with one another. For example, the poles of the earth are tilted 23½ degrees with respect to the poles of the ecliptic. Venus, on the other hand, is only about 3 degrees off-axis, and rotates in the opposite direction, meaning that the up (if defined by planetary spin) for Venus roughly aligns with the down of the solar system. On Venus, the sun rises in the west.

Regardless of your initial perspective, it's clear that if you turn the earth around one full rotation (you may know this as a day), it will look the same as it started. To pick a location on the earth not quite at random: the city of Eureka in Nunavut, Canada (latitude 80° N, longitude 86° W). Residents of Eureka will always find the same heading for north, approximately 1,100 kilometers in a particular direction. Most important, that direction never changes.

The same experiment does *not* work if you somehow shrink a ship down to subatomic size *Fantastic Voyage*–style and park your miniaturized craft above the equivalent spot on an electron. Take out your compass, and there's a good chance that you'll find the up (north) in one direction, but there's a small probability that it's in the opposite direction.

That's quantum mechanics for you. Only certain measurements are possible, and what you might have thought of as impossible turns out to be simply unlikely. And those possible spins have a great deal more to tell us about how a given type of particle functions than we might first have supposed.

---

* I don't actually have a vendetta against Pluto. I just know that there's a very vocal, and surprisingly sensitive, segment of the science world who are still angry about the demotion.

NOT ALL PARTICLES SPIN THE SAME WAY

You may remember Ernest Rutherford's jab about all science being "physics or stamp collecting." Though he couldn't have known it at the turn of the twentieth century, there's a fair amount of stamp collecting in physics as well. It just so happens that the denominations of our stamps correspond very closely to the spins of particles.

Experimentally, each type of particle has a fixed amount of spin, just as each type has a fixed mass and charge. And as with charge, the spins can take on only certain values. What's more, every type particle gets thrown into one of two boxes: bosons and fermions. Just to keep you sane, I've put a little cheat sheet of the fundamental particles at the end of the book. You're welcome.

The simplest particles (at least spinwise) have a spin of 1. For spin, everything is expressed as a multiple of the reduced Planck's constant, written, as we've seen, as a weird little symbol, $\hbar$.

This is an incredibly tiny number. To give you an idea, the second hand on a grandfather clock has an angular momentum about 10^{29} times larger. Spin angular momentum may be small, but it's real. If I were to fire a beam of polarized light—photons are spin-1—at the North Pole, I could eventually stop the rotation of the earth entirely. On the other hand, such a beam would need around 10^{68} photons, a few hundred thousand times the amount the sun will put out in its entire lifetime.

There are a bunch of different spin-1 particles, and they all have something in common. The photon is the mediator for electromagnetism; the gluon is the mediator for the strong force; and the rather unimaginatively named W and Z particles are the mediators for the weak force. Get the pattern?

Collectively, all of the particles that are spin-1 (or any integer) are known as bosons, and they have a lot more in common than just their roles as mediators.

Spin-1 particles are called that because it takes *one* rotation before the particles look the same way as they started. Even though it's going to behave a lot like your monkey brain will think it's *supposed* to work, I don't want you to get hurt patting yourself on the back. Not every particle is spin-1.

There are a few other bosons besides the mediators. There's also the Higgs, about which I'll have a lot more to say in the next chapter, as well as that incredibly elusive particle known as the graviton (maybe). *If* the graviton exists—and lacking a quantum theory of gravity, we don't know that it does—the graviton would be a spin-2 particle. Just as a spin-1 object looks the same after one full rotation, a spin-2 object looks the same after one-half of a turn. This is essentially the same symmetry as grand design spiral galaxies.* You know, like the kind the gang observes at the end of *The Empire Strikes Back*.

Integer spins not only tell us *that* particles act as mediators, they also tell us *how* they act as mediators. Particles with odd spin (photons and gluons and so forth) always produce repulsive forces between particles with like charge. Two electrons, for instance, have the same charge as one another and also repel electrically.

Mediators with an even spin, if they actually exist, behave in the opposite way, in that particles with the same charge will attract one another. Because mass is the equivalent of charge in gravity (a spin-2 mediator) and the masses of all particles are positive or zero, this is just a cool way of showing that gravity is attractive. But of course we already knew that.

But bosons are only half the picture; what about the particles of *you*?

DIRAC, ANTIMATTER, AND FERMIONS

Einstein had shown back in 1905 that matter could be turned into energy and vice versa, but he didn't know exactly *how* to do the alchemy. In 1928, P. A. M. Dirac tried to tidy up the equations of quantum mechanics in a relativistic universe, and in the process, he showed something quite extraordinary: Particles aren't fixed things. Seen at high speeds, electrons can split and multiply into extra particles and antiparticles in a way that is entirely dependent, like the rest of relativity, on perspective.

Because there's no absolute measure of a single electron, Dirac realized that it's insufficient to simply describe where an electron is. Instead,

* My physics lawyer has insisted that I give the following disclaimer. While a spin-1 particle really is identical after one turn and a spin-½ particle is the same after two rotations, it's not actually the case that spin-2 particles are completely identical after two rotations. It's a bit more complicated than that. Also, it would lead to a *giant* digression.

he found a whole pile of related quantum waves, four in total. After a number of false starts, these solutions were interpreted as a spin-up electron, a spin-down electron, a spin-up positron, and a spin-down positron. What he realized is that you can't have one of these without having all of them. Or, more to the point, to properly understand an electron you need to accept that it has many different aspects. It's kind of like Mystique from the *X-Men* in that regard. Or Krishna, depending on your level of cultural sophistication.

This simple superposition tells us more than you might expect. As a reminder, the *quantum* in quantum mechanics means that things like energy or charge or angular momentum can't have any old value that you like. If an electron transitions from one spin state to another, for instance, there will *always* be states that differ by exactly one. There's only one symmetric way to assign spin-up and spin-down so that the difference is 1; make them +½ (for up) and –½ for down.

Electrons are spin-½, but they aren't the only ones. Dirac's equation describes a whole class of particles known as fermions. This includes quarks, positrons, and neutrinos—essentially all of the building blocks of matter and antimatter.

Spin-½ particles are even stranger than they first appear. For instance, while spin-1 particles look the same if you turn them around 1 time, and spin-2 particles look the same after only half a rotation, if you follow the logic with spin-½ particles, it means you have to turn an electron around *twice* before it looks the same as it started.

I realize how ridiculous this sounds. After all, when I say, "Turn an electron around," what I really mean is that we should turn the entire universe around 1 full rotation, and because, by definition, a full rotation should bring the universe exactly to where we started, there's clearly nothing in your classical arsenal to deal with this.

That's okay. We're not dealing with the classical universe. In the last chapter, we saw that there are two things you can do to the wave function of the universe to make everything "look the same." One, as you might expect, is to multiply by 1. The other, surprisingly, is to multiply by -1. A full rotation of a fermion gives you a -1, and a second rotation gives you another -1. Multiply them together and even at the quantum level, everything returns to where you started. We saw a similar set of options at the end of the last chapter.

As Werner Heisenberg put it, rather poetically:

> We have to remember that what we observe is not nature herself, but nature exposed to our method of questioning.

There is a little experiment we can do to convince our brains that sometimes the universe isn't quite as obvious as it might seem at first.

This particular version of the experiment is usually called Feynman's Plate, but there are lots of others, including Dirac's Belt, and the Quaternionic Handshake. You can probably get pretty good versions of the others by looking on YouTube.

For my version, I want you to start with a cup full of water in your

Feynman's Plate

FIRST ROTATION

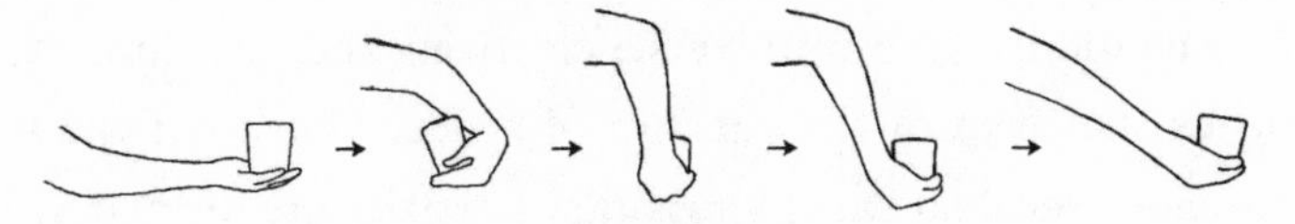

SECOND ROTATION

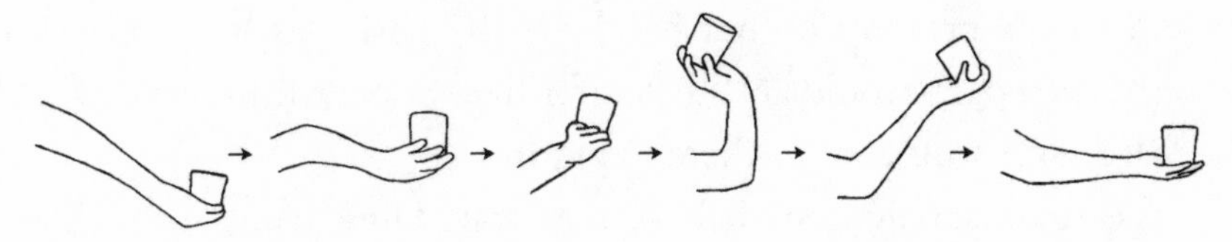

right hand, palm up. The water is important, because if you screw up the experiment, you'll be quickly informed by a lap full of water.

Rotating from your elbow, turn your arm underneath your shoulder one full turn. Be careful not to spill any water. You've now turned 360 degrees, but as you've probably noticed, your arm is *not* how you started. Your elbow should be pointing straight up, somewhat uncomfortably.

Finally, rotate your arm *another* full turn, in the same counterclockwise direction as the first time, but this time bringing your cup and wrist *above* your shoulder.

Tada! If you've done everything correctly, you should (1) find your arm and cup in precisely the same configuration as you started and (2) not have spilled any water on yourself. More to the point, it took you two full turns to get there.

SYMMETRY AND ANTISYMMETRY

Electrons are microscopic gyroscopes that somehow, through the power of mathemagic, look the same only if you turn them around twice. These features, however, are just a curiosity compared to what comes next.

When Dirac first derived the properties of spin-½ particles, he concluded something interesting—but wrong. The antiparticles, he thought, jump out of the equations with an apparently negative energy.

I want to give you a sense of just how catastrophically unstable such a universe would be. In our universe, electrons and positrons constantly pop into existence from the vacuum of space. That's not the big deal. We've already seen that these temporary particles are ultimately responsible for the evaporation of black holes.

On the other hand, if those particles *weren't* temporary, then we would have a problem on our hands. If the positrons had a negative energy, then pairs would get created for free. Soon, the universe would be filled with stuff and nowhere to put it.

Electrons simply don't behave that way. Dirac came up with his famous equation in 1928, but 2 years earlier, before all of this noise about spin and antimatter was even in the air, the Italian physicist Enrico Fermi—for whom, you'll be unsurprised to learn, fermions are named—found that electrons behave in a fundamentally different way from particles such as photons.

Photons tend to cluster at very similar energies and phases (which is how lasers work), while electrons, especially electrons at low energies, tend to settle into different states. This is why bosons (such as photons) and fermions (such as electrons) were sorted into different piles long before we really understood the symmetry relations that described their different behavior.

It was only later that we learned that the differences between the two groups are ultimately due to the different spins of the particles. In 1940, Wolfgang Pauli formalized this difference with the Spin-Statistics Theorem. Pauli's theorem essentially says that whether a particle is a boson or a fermion entirely determines what happens when you swap identical particles in a system.

To understand how this works, we need to return to the wave nature of quantum mechanics. Quantum waves have a few important features: amplitude, frequency, wavelength, and so on. But one of the most overlooked features of a wave is something known as the phase. At any time and place, a wave can have a positive or negative value, depending on whether you're nearer to a peak than to a trough; we don't really care which, however, because the only thing that comes into play is the *square* of the wave. But as the wave propagates, it oscillates between the two. The phase is just a number that describes the timing.

Think of a shift in phase like a musical round. You sing "Row, row, row," and I come in a few beats later. It's only the *difference* in our phases that makes the round work.

Quantum mechanics takes this a step further. In quantum mechanics, not only can't you measure the phase of a wave but the phase doesn't even have any definite value. If it did, you'd have all of the information

Wave Interference

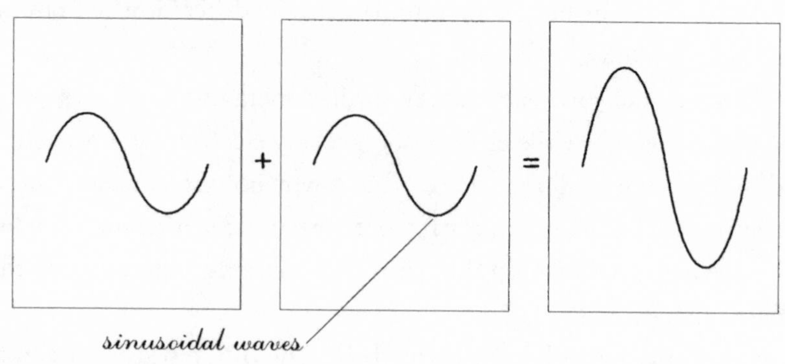

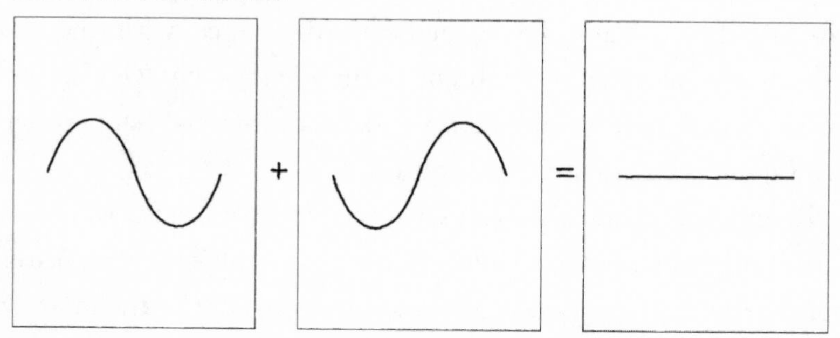

necessary to do quantum teleportation without all of that messy entanglement that we talked about earlier. But we *do* care if two waves are in phase or out of phase. In the former, the two waves add constructively, making the probabilities larger than they would have been otherwise. With the latter, they partially cancel. This is how you can get acoustical dead spots in a symphony hall.*

Even though we don't get to measure phase directly, it's a good proxy for what happens when you rotate a particle. If you rotate a fermion just

* Who am I kidding? We're all philistines here, right?

once, its wave function will be 180 degrees out of phase from where it started. This is exactly equivalent to multiplying by a minus one.

Rotate it a second time, and there's a 360-degree change in phase, which amounts to no change at all. This is exactly the rule for a spin-½ particle.

I'm about to pull a fast one. You see, the minus one that shows up when you rotate a fermion (and correspondingly, the plus one that shows up when you rotate a boson) is exactly the same as the minus one that shows up if you *swap* two fermions of the same type.

The numbers (-1 for fermions, +1 for bosons) are the same, and the mathematics of rotation and replacement are closely related, but beyond that, I'm afraid that there's no simple analogy for *why* these two phenomena should have the same factor. As Richard Feynman rather exasperatingly put it:

> We apologize for the fact that we cannot give you an elementary explanation. . . . It appears to be one of few places in physics where there is a rule which can be stated very simply, but for which no one has found a simple and easy explanation. . . . This probably means that we do not have a complete understanding of the fundamental principle involved.

All of this was building up to a minus one. Fermions have it; bosons don't. It's a very simple symmetry.

Looking at the two options, minus one and plus one, gives us a *very* good idea why there are only two groups of particles and gives a hint as to why they behave so differently from one another.

WHY −1 IS SO IMPORTANT

Negative one shows up *all the time* in physics, most obviously in electric charge.

It's an accident of history—a convention due to Ben Franklin—that electrons have a charge of –1, while protons have a charge of +1. Franklin figured that when you rubbed wool on wax, the wool ended up with extra electricity and the wax ended up with a deficit. Later, it became clear that the electrons from the wool were being left on the wax, essentially giving rise to the choice of signs. Truth to be told, though, it doesn't matter which is which, so long as we're consistent. The only thing that determines the interaction is the product of two charges. A negative times a negative is a positive, which means that the product of two electrons and two protons are the same: +1. That plus sign means that like charges repel one another.

We get the same sort of relation when talking about fermions and bosons. Fermions, as we've seen, get –1 and bosons get +1 when you swap two identical particles. We don't even need to worry about swapping, just about what number goes with what kind of particle. You can even figure out how a composite particle behaves by simply multiplying out all of the components. There are three quarks (fermions) in a proton, and if you multiply –1 by itself 3 times, you end up with –1 again, and *that* is why protons and neutrons are fermions.

Or to give yet another example that I glossed over in Chapter 1, some particles look the same in the mirror, and some look reversed. This describes a property called parity, a property that is every bit as fixed as the charge of a particle. A quark, for instance, has a parity of +1, which means that the version in the mirror is the same as the original. Just like with charge, the antiparticle will have the opposite sign; the antiquark gets a parity of –1.

Parity works the same way as the particle exchange rules. If you've got more than one particle, you just multiply them together. Your hands, for example, are both antisymmetric (parity of –1)—they appear reversed in the mirror—but so long as you reflect both of them, the overall reflection looks symmetric (parity of +1).

A pion is a particle made of a quark (+1) and an antiquark (–1). I can figure out how it will look in a mirror by just multiplying the parities of the components. A pion has a parity of –1.

By the exact same rules, a pair of pions has a parity of +1, and *three* pions have a total parity of –1. All of this may seem like a curiosity until you realize that parity is *supposed* to be conserved, and it absolutely is in electromagnetic and strong reactions.

But weak reactions are another matter entirely, and this is the key to *really* understanding the Cronin and Fitch experiment that we talked about earlier in the book. As you'll recall, Cronin and Fitch were studying kaons in 1964. Kaons are weird little beasties.

Make a bunch of kaons, and about half of them will survive for only a billionth of a second or so. The other half typically lives about 600 times longer. This is such a huge difference that the long and short kaons are effectively different particles.

They typically leave behind very different corpses when they decay. Both versions decay into pions, but the short-lived version decays into two pions (parity of +1), and the long-lived version decays into three (parity of –1). Before Cronin and Fitch, everyone assumed that *all* kaons should follow this pattern.

They don't. About 1 in 500 long-lived kaons decayed into products that were supposed to be completely impossible: two pions. Some long-lived kaons essentially turned into short-lived ones. The difference isn't a small one. A conserved quantity—the parity—apparently isn't conserved after all, and this, as we've seen, is a necessary prerequisite for all of the matter in the universe. But really understanding why understands a mastery of the symmetry of –1.

I've spent an enormous amount of ink trying to motivate the difference between different kinds of particles, and at the end of the day, all of those differences result in a minus sign that you can't even measure.

I can read your mind, and it sounds a little like this: "Who cares?"

THE PAULI EXCLUSION PRINCIPLE

Let's recap this line of reasoning.

> We live in a quantum universe where the square of a quantum mechanical wave tells you where you are likely to find a particle.
>
> Some particles, the fermions, put a minus sign in front of those waves if you swap the positions of two of them.
>
> None of this apparently matters because the probability is the square of the wave, so the minus sign gets multiplied away.

Have I mentioned that you owe your existence to that minus sign?

The particles of matter (including us) are fermions: protons, neutrons, quarks, and electrons. Imagine two electrons with their spins in the same direction, one exactly on top of another (and for those experts reading this, they both have the same momentum as well). Now imagine swapping them. To us, nothing seems to have changed, but to quantum mechanics, the whole universe is now in chaos. The wave function shouldn't change at all because there's no distinguishing one electron from another, but somehow we're supposed to put a minus sign in front.

To recap: Change nothing but multiply by minus one. The only number this works for is zero. In other words, zero wave function, so zero probability, or no chance whatsoever. It can't happen.

You can't get two electrons in exactly the same place at exactly the same time with the same spin. When Pauli discovered this in 1925, he hit on the key to understanding things as diverse as the electronic structure of atoms all the way up to the final fates of stars. This is known, quite rightly, as the Pauli Exclusion Principle.

I cannot overstate what a big deal the Exclusion Principle is. To the uninitiated, it might sound like a curiosity. You probably came into this book assuming that particles can't overlap one another anyway, so all of the mental gymnastics of switching particles and minus signs might have seemed kind of pointless. What possible use can the Exclusion Principle have on your everyday life?

Atoms—and thus you, me, and any aliens that might be out there—are made of fermions from top to bottom. To pick two elements not entirely at random, the main difference between hydrogen and helium is that hydrogen has one electron and helium has two, each to match and neutralize their one and two protons, respectively. That minor difference makes a world of difference once you introduce the Exclusion Principle.

We've spent a lot of time talking about spin of an electron, but throughout it all, I've stressed that there are only two possibilities: up and down. Because of the Exclusion Principle, an atom can get only two electrons in the lowest energy state, one for spin-up and one for spin-down. An ordinary neutral hydrogen atom doesn't especially *want* to accept another electron, but it's not forbidden either. You could, if you like, shove another electron in there, and your hydrogen ion would happily bond to other, positively ionized atoms. It's this accommodation that allows hydrogen to bond with oxygen to make water or with carbon to make methane or any of the hundreds of other chemicals that have hydrogen in them.

Helium doesn't have this luxury; it has a full house. And unlike a literal full house in which someone could, if they really had to, sleep in the bathtub, helium is completely out of options. There are only two possible spin states and the ground state is full, so it *cannot* accept any more electrons.

Poor helium is almost always on its own. Its lack of chemical reactivity means that helium balloons are far safer than their hydrogen counterparts. The same is true for neon, argon, and the other so-called noble gases.* In each case, only so many electrons will fit into various shells of electrons. The details are a little complicated (and not entirely necessary), but the lowest level fits only two electrons (helium), the next level fits another eight, for a total of ten (neon), and so on. The noble gases keep their electrons to themselves; they share with *nobody.*

On the other hand, elements with lots of extra electrons outside those shells—things like copper and gold and so on—are excellent electrical conductors; they are the free-loving hippies of the atomic world, perfectly happy to share their electrons.

All of the particles in your everyday life—neutrons, protons, and electrons—obey the Pauli Exclusion Principle because they are fermions, and that exclusion makes all of the good things in the universe possible. But, as we've seen, not *every* particle is a fermion.

The world of bosons is a very different place. Because of their plus sign, bosons couldn't care less about the Pauli Exclusion Principle. If you take certain materials, helium nuclei, for example, or photons, and cool them down to incredibly low temperatures, you'll find that something surprising happens; they condense into frictionless superfluids, which means that they can do things like flow without viscosity, form into ultrathin layers, and even seemingly defy gravity in an effort to reach minimum energy.

It's also a good thing that photons are bosons, because otherwise we wouldn't have lasers. The magic of lasers is that all of the light is perfectly synchronized in exactly the same energy and the same state. This wouldn't be possible with fermions.

Bosons are fine for lasers, but they are very bad for us. Without fermions, we don't get any of the chemistry or structure that we see in our everyday world. In other words, but for the existence of a minus one, you wouldn't be here; the universe would be about as unanthropic as you can get.

* I've always found this naming a bit patronizing.

WHITE DWARVES, NEUTRON STARS, AND DEGENERACY

The Exclusion Principle* is necessary for existence in more ways than one. You are, as Carl Sagan famously put it, the "stuff of stars." At first glance, we might suppose that the Exclusion Principle doesn't have much to do with stars because it kicks in only when two particles are in danger of overlapping with one another (physicists rather judgmentally refer to this as *degeneracy*). In stars, this normally isn't much of an issue because the high stellar temperatures do a good job of propping up a star, using pressure.

Stars are the ultimate laboratories in which the complex atoms that make you were formed. The Big Bang made a bunch of simple stuff: hydrogen and helium, primarily; but to get heavier elements, you need stars. Our sun fuses hydrogen into helium. In the distant future, it will run out of hydrogen and will be forced to subsist on a more meager nuclear diet of fusing helium into carbon and oxygen. Ultimately, however, it will run out of nuclear fuel entirely and will begin to collapse under its own weight. The sun will then smolder more or less eternally as a white dwarf. Don't worry; we've got another 5 billion years or so yet.

As a white dwarf, the sun will butt heads with the Pauli Exclusion Principle. Once it runs out of fuel, the sun will start to cool. And, like a hot air balloon, when it cools, it will collapse. Here's where things get weird. As it continues to collapse and shrink—ultimately to the approximate size of the earth—the gravity on the sun will get stronger and stronger.

But that's when the Exclusion Principle intervenes. There are electrons flying around, remember, and as the sun collapses, those electrons

* And, lest you forget, the Exclusion Principle is really just a natural consequence of the fact that you can replace a particle with another of the same type without God noticing.

will ultimately be packed tightly. Really tightly. The fundamental physical laws of the universe—informed by Particle Replacement Symmetry—prevent two electrons from overlapping one another. The collapse will grind to a halt, and the whole thing will literally be held up by uncertainty.

The only reason that electrons can't be squeezed so tight that they are overlapping is that they are spin-½ particles, and we live in a universe of Replacement Symmetry.

Degeneracy tells us not only about our fate but about our origins as well. I mentioned that our own star makes helium, oxygen, and carbon. Somewhat more massive stars can also put out heavier stuff, such as neon, magnesium, silicon, and iron. But there's quite a lot of material lying around on earth and elsewhere that isn't made of any of that.

Where did everything else come from?

To answer that, we need to look to the most massive stars. Eventually, no matter how large a nuclear gas tank you start with, a star is going to run out of fuel. In fact, even though they have more hydrogen to burn than the sun, the heaviest stars burn *much* hotter and *much, much* brighter. As a result, these stellar heavyweights live fast and die young.

The most massive stars burn through everything they can, but ultimately, they're left with iron. This is the end of the line. No matter how much you squeeze, it takes more energy to fuse iron into something heavier than you get out in the process. That's why nuclear *fission* always involves uranium or plutonium and the like—materials far heavier than iron. You get energy out by ripping them apart. But poor, lonely iron is just the slag heap of the nuclear universe.

Without nuclear burning, the heaviest stars have no pressure and nothing to hold them up. Unlike the sun, however, electron degeneracy pressure isn't enough to hold the most massive stars up. Using the enormous gravitational energies at its disposal, the star shoves all of the protons and electrons into one another and turns the whole thing into neutrons.

I bring up neutrons because there's a weird quirk with uncertainty. Heavier particles can be squeezed tighter than can lighter ones before the Pauli Exclusion Principle kicks in. Neutrons are about 2,000 times the mass of electrons, so the newly formed neutron star can collapse about 2,000 times smaller than a white dwarf of the same mass. Though the term *neutron star* seems innocuous enough, beware. They are two to three times the mass of the sun, but they are only about 5 kilometers in radius. Tiny, compared to a normal star.

Now picture this: You have the incredibly rigid, incompressible core of a star confined to a tiny, tiny region of space. All of the material in the outer regions of these stars suddenly notice, à la Wile E. Coyote, that the floor has been pulled from under them. That gas, a goodly fraction of a star's worth, starts to fall, and by the time it hits the neutron star core, it's traveling very close to the speed of light.

And then it bounces.

It slams into all of the other in-falling material and produces one hell of an explosion, one that can be seen from neighboring galaxies. You may know it as a supernova.

Supernovas do two important things from the point of view of making our existence possible. First, they distribute energy and some of the lighter elements out into the Galaxy: carbon, nitrogen, oxygen, iron, and so forth. These are some of the most abundant elements around, and if you chop off your arm and throw it into a mass spectrometer, you'll notice that you are mostly made of stellar detritus.

But some of the heaviest elements can't be made in stars at all. We saw earlier that a star can fuse only elements lighter than iron and get energy out. As result, anything *heavier* than iron has to be made some other way, and that other way is supernova explosions. Nickel, copper, gold, and (Superman's own) krypton are some of the many elements that would be energetically impossible without supernova explosions.

We are the result of one such cataclysm or, mostly likely, a couple of them. Look around you, and you see a world filled with heavy metals. Some of them are used in our tools, and some of them are used in us.

WHAT WOULD A TEASPOONFUL OF NEUTRON STAR DO TO YOU?

It's easy to forget how we got here. Neutron stars—and thus supernova explosions and thus heavy metals—are ultimately based on Particle Replacement Symmetry. After all, it was Particle Replacement Symmetry and the Rotational Symmetry of fermions that gave rise to the Pauli Exclusion Principle, which in turn gave rise to degeneracy pressure.

What a long and winding road to take a look at some of the strangest occupants of our universe: neutron stars. If nothing else, they serve as a handy reminder of the extreme emptiness of space under *normal* circumstances and the extreme power of a simple symmetry.

Neutron stars are all around us, and though they seem fairly small, they are extremely dangerous. Because they are perhaps the best exemplars of the Exclusion Principle, I wanted to give you an idea of how seriously dense something has to be before degeneracy kicks in. As a thought experiment, suppose you wanted to get a teaspoon of the stuff and bring it back home. What would happen?

Warning: DO NOT TRY THIS AT HOME.

Because of this unbelievable density of a neutron star, the gravity is incredibly high. You might well expect the entire thing to collapse into a black hole, and you'd be very nearly right. This is precisely the reason that neutron stars can't be more than a few times the mass of the sun. Otherwise, they *would* become black holes.

So what would happen if you were fool enough to approach one of these things?

Landing is going to be incredibly tough. Neutron stars can spin at thousands of times per second and many of them have magnetic fields over 10 million times stronger than the earth's. This is going to adversely affect you in a few ways. First, magnetic fields at those levels are almost certainly going to destroy *anything* with ferromagnetic materials (a fancy word for things like iron that you can make magnets out of) as well your computer systems.

Also, the combination of spinning and strong magnetic fields means that neutron stars essentially have their own defense system. You may know them as *pulsars* and they basically consist of a high-energy radiation beam sweeping through the sky every fraction of a second. Finally, have you ever tried to land on a planet whose surface is rotating at thousands of kilometers a second? Try writing that science fiction story. It isn't easy.

But supposing you could land on the surface of the neutron star. Sure, it's something like a million degrees Kelvin, but compared to the other problems you're likely to encounter, that's child's play. The gravity is something like 200 billion times that on the surface of the earth. If that doesn't concern you, consider that the *difference* in gravity between your head and your feet is approximately 60 million *g*. You would be squashed almost instantly.

Because I like you, I'll allow you to survive for a bit longer. We've already discussed the design specs for a *Star Trek*–style transporter device, so we might as well make use of it. Let's suppose you transport up a teaspoon of neutron star from the core directly into the cargo bay. I say "the core" because the outer crust is kind of boring; it's mostly heavy elements like iron. To get to the pure product, you need to dig deeper.

What happens next? Here's where the fun *really* starts.

You first need to realize that we're talking about densities of about 10^{18} kilograms per cubic meter, which means that the total mass in a teaspoon is somewhere on the order of 10 billion tons. That's about the mass of a reasonable-size mountain.

Inside a neutron star, there's a delicate balance between the tremendous gravity of the star and the degeneracy pressure of the neutrons. Once we extract the neutrons, all bets are off. We no longer have the gravitational pressure to compress our neutrons together, and remember, these neutrons are at temperatures of millions degrees. The gas pressure is *huge*. Even if you could use a transporter to teleport your neutron star into the hold of your ship, the sudden decrease in pressure will cause the gas to explosively expand. Assuming a decent-size cargo hold,* your teaspoon of pure neutron goodness will end up producing a pressure of something like a quadrillion times atmospheric normal and a density on the order of 10 million times that of solid rock.

Do not stand in your cargo bay when you beam up your neutron star material. I cannot stress this enough.

Supposing the expansion of neutrons didn't destroy your ship outright, the worst is yet to come. Inside a neutron star, the degeneracy pressure also stops neutrons from doing what they'd normally want to do: decay. Neutrons can hang out for a very long time if they're inside atomic nuclei, but on their own, they're not terribly long lived, at least on normal human timescales. Compared to many subatomic particles that last only a billionth of a second or even less, the 10-minute lifetime of a neutron is incredibly long. After that 10 minutes (on average) a neutron decays into a proton, an electron, and normally undetectable antineutrino.

Not a big deal, right? Wrong. We finally get to invoke $E = mc^2$. In the case of neutron decay, about 0.08 percent of the mass gets converted to energy in the process, which doesn't sound like too much, but multiply it over your teaspoon of neutron star, and it ends up producing an equivalent energy of roughly that put out by the sun every 2 or 3 seconds.

The neutron decays release the equivalent energy of a trillion megaton nuclear device. To put things in perspective, the first nuclear bombs were about 200 *kilo*tons. Your teaspoon of neutrons would easily be enough to destroy all life on earth. Congratulations, you've teleported a live nuclear device onto your ship.

Best of luck to you.

---

* I've used the online design specs for the *Enterprise-D*. Don't judge me.

Chapter 9

HIGGS

IN WHICH WE EXPLORE THE ORIGIN OF MASS AND WHY PHYSICS *ISN'T* STAMP COLLECTING

We can observe galaxies billions of light-years away and detect microscopic particles with lifetimes of tiny fractions of a second. We can break the pull of the earth's gravity and are mere months away from developing wedgie-proof underwear.

I think science has done a pretty damn good job, don't you?

But still, science can be kind of *list-y* at times. There's a famous story wherein Enrico Fermi (who won the Nobel Prize in 1938) was speaking to his student Leon Lederman (who himself won the prize in 1988).* Lederman asked Fermi about some particular particle or other, to which Fermi replied:

* Lederman also wrote a very nice book on symmetry and, in another book, coined the unfortunately sticky—and somewhat misleading—appellation the *God Particle* for the Higgs boson. I implore you—please don't call it the God Particle.

> Young man, if I could remember the names of these particles, I would have been a botanist.*

Particle physics, much like the Industrial Revolution, progressed by noting that big, complex things are made of smaller things made from interchangeable parts. Literally millions of different molecules can be built out of only a few hundred different atomic isotopes. Those atoms are, in the end, made of only three particles: protons, neutrons, and electrons. How deep does the rabbit hole go?

It would be fantastic if the universe really were built up from just three particles, but for some reason, there are lots of "elementary" particles that don't seem to do much of anything. There are at least twelve different fermions, and at least five different types of bosons, each with different spin states, antiparticles, and so on, giving a grand total of sixty-one. This is to say nothing of the literally hundreds of different composite particles. We have a laundry list of particles and forces but, so far in our story, no real idea of where they come from.

That is about to change.

The complete collection of particles and forces has come to be known as the Standard Model. It is one of the great triumphs of human thought, and one, you'll be pleased to know, that goes far beyond simply listing particles and properties.

There's a very deep structure to all of it. The particles, the forces—all arise out of symmetries.

NOTHING IS (REALLY) REAL

You are living in the *Matrix*. While everything around you seems substantial and permanent, in a very real sense, they are neither. Particles,

* Which, I think you'll agree, is a much more respectful version of Rutherford's comment about all nonphysics sciences being essentially stamp collecting.

even elementary particles, aren't fixed and constant things. We've seen that in the vacuum of space, particles and antiparticles are created from nothing and return almost as quickly.

In a universe of quantum uncertainty, it doesn't make much sense to talk about a single particle. Think, instead, of a vast, unrelenting *swarm* of electrons flying through the universe en masse. Taken as a whole, or quantum mechanically, it's hard to say for certain where one electron ends and another begins. So we don't. Instead, we describe the electron-ness of the universe as a field.

The universe, even when it's apparently empty, is filled with fields. Though you've heard the word in the colloquial sense of the term, a field is something special to a physicist. Think of fields as you would the Force. "It surrounds us and penetrates us; it binds the galaxy together."* We've seen a few fields already. Maxwell's great contribution was describing electromagnetic fields. As Einstein put it:

> Before Maxwell people conceived of physical reality—in so far as it is supposed to represent events in nature—as material points, whose changes consist exclusively of motions, which are subject to total differential equations. After Maxwell they conceived physical reality as represented by continuous fields, not mechanically explicable. . . . This change in the conception of reality is the most profound and fruitful one that has come to physics since Newton.

Einstein, famously, showed that the electromagnetic field wasn't some indivisible goop, but rather could be broken down into discrete particles. You know them as photons. Just as water molecules and water waves are two sides of the same coin, photons and electromagnetic

* Technically, it's gravity that binds the galaxy together. Fortunately, gravity *is* a field.

fields are just two different ways of thinking about the same thing.* How we look at it really depends on the context. With visible light—the kind our eyes are sensitive to—we can count individual photons. For longer wavelengths, it's radio *waves*. But at a fundamental level, they're the same thing.

There's a gravitational field too, and there are fields for the other fundamental forces. These sorts of fields are common enough to enter the vernacular. On the other hand, you rarely hear someone talking about an "electron field." And yet, you're soaking in it.

Every particle has a corresponding field. At the simplest level, a field tells you how many particles are in some region of space and how fast they're going.

But if you want to be more sophisticated about it, you can think of a field as a trampoline with a bunch of hyperactive children jumping on top. With every jump, little ripples will emanate outward. Keep your eyes on them. If we have the math to interpret them, those ripples can tell us everything we want to know about particles zipping through the universe: the density of particles, their momentum, and so forth.

Practically speaking, what jumping up and down looks like—in the case of electromagnetism, for example—is wiggling an electron around. This is how a radio transmitter works.

But the analogy isn't perfect. A trampoline is a two-dimensional surface and we live in three-dimensional space. If you can accurately picture a three-dimensional trampoline bouncing around in the fourth dimension, congratulations, you are a Borg.

---

* A note to aspiring experts. You're going to want to think of fields and quantum waves as being the same thing. They're not. Although they have a lot of properties in common (interference, frequency, and other wave properties), a quantum wave describes a single particle, whereas a classical field describes the ensemble properties of a whole mess of particles.

Fields and Particles

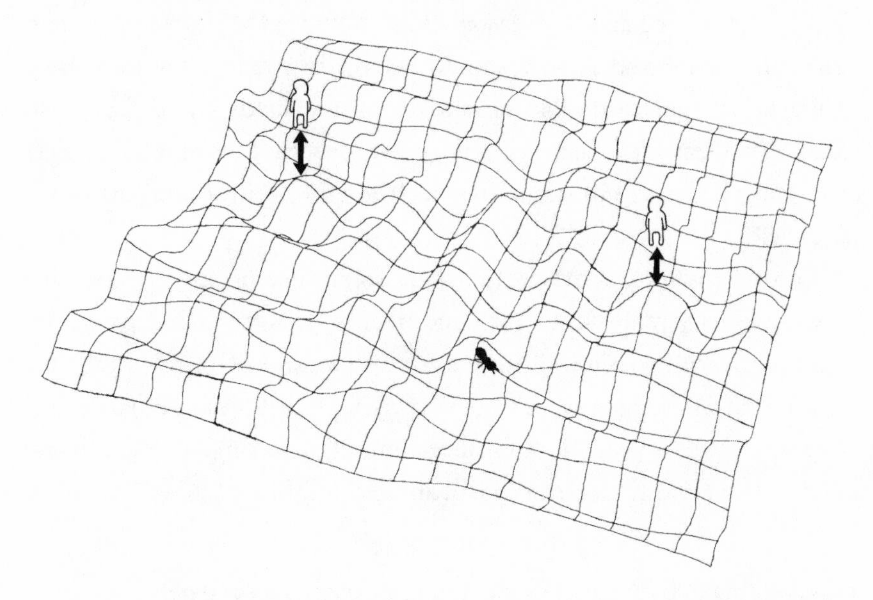

Throw some pebbles into a pond, and you'll see that you can get some *very* complicated patterns with nothing more than the interference of the resulting waves. You don't even need to imagine. Everything you see is nothing more than a collection of an enormous number of electromagnetic waves stacked on one another and beamed into your eye. Sound works the same way except, obviously, on your ears.

At first blush, it may not seem as though the concept of a field is especially important. Instead of dozens of different fundamental particles, we're just replacing them with dozens of different fundamental fields. Good job, physics!

But the world of fields is going to reveal symmetries that particles never could. A few simple fields are going to suffice for describing virtually all of the particles in the universe.

Naively, you might have supposed that we're going to need sixty-one different fields, one for every possible type of particle, color, spin, and so forth. Sixty-one fields seems like an awful lot of work, especially if we need to come up with a different rule governing each one.

On the other hand, it's reasonable to suppose that left-handed particles and right-handed ones should behave in roughly the same way, so maybe we get to double up and use some fields twice. Likewise, recall Wheeler's idea that a positron looks exactly the same as an electron going backward in time. The electron and positron are part of the same field.

In other words, some particles are so similar to one another that they are clearly made from the same recipe. Electrons and positrons, spin-up and spin-down, all have a lot on common—mass, total spin, the amount of charge—so it's not surprising that they're all part of the same field. We can get a lot of mileage out of exploring what a universe filled with only our electron–positron field might look like.

HOW CHARGE COMES FROM SYMMETRY

A universe with just electrons and positrons is an incredibly boring place. There'd be no molecules, of course—not even any atoms. Your team of kids could metaphorically jump up and down on the cosmic trampoline to their hearts' content, generating electrons and positrons, and the waves would just pass through one another. Without photons, charged particles have no way of interacting. To create interactions, we're going to have to delve into the world of *internal* symmetries.

For much of this book, we've talked about symmetries that arise when we fly around in a spaceship or look at things in a mirror. There is a direct analog to your everyday experience, in that you can look in a mirror or fly in your spaceship and verify that I haven't been making all of this up. With fields, those external symmetries are important, but internal symmetries are even *more* important.

Internal symmetries are incredibly subtle, and rather than give a definition, let me give you an example. In the last chapter we saw that waves have a hidden quantity called phase, hidden because we can't ever measure it directly. *Nothing* changes if you alter the phase. You will be unsurprised to learn that phase is an internal symmetry.

Bounce up and down on the cosmic trampoline with some fixed frequency. As you jump, waves of electrons and positrons emanate outward. Lesser creatures living on the surface of the trampoline (our ants from Chapter 7) notice an average fluctuation, and their lab equipment tells them that this corresponds to a flow of electrons and positrons through their labs.

Adjusting the phase of the bounces is super easy. All you need to do is time your jumps a fraction of a second earlier or a fraction of a second later. To the ants, a change of phase has absolutely no detectable effect. They still see the same flow of electrons.

Or because the theme is physics, let's talk about magnets and in particular the one underneath your feet. The earth is a giant magnet, as you certainly know if you've ever used a compass. Strangely, though, every few hundred thousand years, the north magnetic pole and the south switch.

You could imagine living through the epoch of reversal, and while things might be a bit confusing for a while, all you'd have to do to navigate is relabel your compass and you're good to go. What makes everything so simple is that in one fell swoop (technically a fell swoop that probably takes a few thousand years, but bear with me) *all* compasses would reverse themselves.

This is what we call in the biz a global symmetry transformation (*global* because you're relabeling everything in the same way) and a pretty simple one, at that. Whether your compass needle points north or south, you can still reliably use it to walk a straight line.

The mathematics of phase is almost identical to compass needles. Both phases and compasses loop around again once you reach a certain point. Imagine turning a dial and for every little turn, the phase shifts just a little bit. But once you've gone around full circle, you're back to where you started. This is the simplest and most fundamental of symmetries: the symmetry of a circle.

> **Global Phase Symmetry:** The phase of a field can change everywhere in space and time by a fixed amount and there will be no measurable effect.

Mathematicians classify symmetries using their own special language called group theory. They don't get overly concerned about whether we're talking about the symmetry of a quantum system, the direction of a compass, or the phase of a quantum field. The name they give to the Phase Symmetry is U(1).* It looks frightening, but the takeaway is that the *1* means there's a single number—the phase itself—that can be changed without anyone noticing.

Were it not for Emmy Noether, this might have been only a curiosity, but Noether told us that if there's a symmetry, even an internal one like phase, then there's going to be a conserved quantity. In this case—and I'm not actually proving this, mind you—we get conservation of electric charge.

> Phase Symmetry → Conservation of charge

In the beginning of the book, I said that conservation of electric charge was a thing enforced by the rules of the universe, but now we see this is *why* it's conserved. It's just a simple result of Noether's Theorem and Phase Symmetry.

---

* Except for a helpful lookup table you'll see in a bit, this is the first and last time I'm going to use the technical group theory version of this symmetry. It is *entirely* for impressing people at parties.

WHY THE UNIVERSE NEEDS PHOTONS

Phase Symmetry does more than simply give us conservation of charge, and to understand why, we need to look locally.

Imagine the Mad Hatter wanted to play a trick on (the fairly literally minded) Alice. He hands her a compass and tells her to walk in a straight line a certain number of paces. Alice looks at the compass, and so long as the needle points in the same direction the whole time, she can be sure that she's walking in a straight line.

Because he's mad and clearly at least a little bit mischievous, the Hatter might plant a series of magnets in the nearby terrain. This has the effect of adjusting the orientation of the compass needle by different amounts in different places. Alice, following the directions of the needle, takes an incredibly erratic path through Wonderland, all the while thinking that she's walking a straight line.

Even though Alice is still going "directly" according to the rules following a compass, anyone watching from the outside would see that there is clearly another influence at work.

In the language of physics, the Mad Hatter is performing a *local transformation*.

And now for the symmetry. We're going to *assume* (and this turns out to be a well-founded assumption) that no matter how the Hatter places his magnets, Alice's motions will be something predicted by physics.

The idea is that you could go in by hand and adjust the phases of an electron field by a different amount at every point in space and time. Even though the dynamics of the field would become far more complex, the assumption is that those dynamics are valid. All you have to do is introduce another force at work—in our case, some hidden magnets.

This is an odd sort of symmetry. It goes by the name of *Gauge*

*Symmetry**, and if it didn't actually work, you'd be crazy for thinking that our universe is invariant under Gauge Symmetry transformations. We, or really Hermann Weyl, who was the first to introduce them, use them because they work. They end up producing the physical laws that we see.

Gauge Symmetry supposes that there needs to be *another* field besides the electron–positron field to explain all of the interactions and interferences. And just like Alice walking around in a wandering path, the source of all of this apparent chaos can be explained through something external: electromagnetism. In other words, if we want to restore symmetry, we need to introduce electromagnetism and, with it, the photon field.

In the late 1940s, Sin-Itiro Tomonaga, Julian Schwinger, and Richard Feynman realized they could essentially *derive* all of electromagnetism from scratch if they simply assumed that phase was a Gauge Symmetry. But to make the symmetries work, they needed to add two additional components into the equations:†

1. The equations of motion describing the photon field.
2. Interaction energy between photons and charged particles.

Everything simply pops out *as if by magic.* The equations reproduce all of Maxwell's equations essentially from first principles. They predict that the photon must be a completely massless spin-1 particle, predictions that are perfectly borne out by experiment.

Technically, these are all *post*dictions. We knew what photons and

* The etymology is a bit murky, but the gauges of gauge transformations are meant to bring to mind the scales and timing of railroads.

† When I say *our equations*, what I really mean is "the Lagrangian of the universe." The Lagrangian, as you'll recall from Chapter 4, describes all of the interaction energies, and is the thing that needs to be symmetric.

electromagnetism were supposed to look like before we started any of this. Nevertheless, the beauty of the symmetry approach is that we get everything—literally all of the laws of electromagnetism—from a simple assumption about symmetry. The *only* missing piece is that the charge of the electron, the strength that charged particles interacted with the electromagnetic field, has to be put in by hand. This theory is incredibly elegant, but just to get your Spidey senses tingling, *any* time a theory has a number that has to be manually adjusted, that's a pretty good indication that we're not at the end of the story.

WHAT *REALLY* MAKES TWO DIFFERENT KINDS OF PARTICLES?

We got really lucky with electromagnetism.

Maxwell gave us his equations back in the nineteenth century, and although reframing everything in terms of Phase Symmetry was a huge intellectual accomplishment, it's a hell of a lot easier to figure out a problem when you already know the answer. Even so, it was a lot more than math for math's sake; it motivated the idea that symmetries might generate other forces. (Spoiler alert: They do.)

In 1954, Chen Ning Yang and Robert L. Mills, both at Brookhaven at the time, provided a general process for turning symmetries into forces. Yang and Mills were the intellectual heirs to Emmy Noether, and they took her obsession with symmetries and invariants to Escheresque extremes.

Noether, remember, said that if you have a symmetry, you get a conserved quantity. Yang and Mills said that if you assume a *Gauge* Symmetry—like the trick with placing magnets to screw up a compass—then there must be one or more mediating particles. In other words, symmetry doesn't just give us conservation laws. According to Yang and Mills, if you assume a symmetry, you get a fundamental interaction from whole cloth.

This is harder than it sounds.* Mathematicians have piles of symmetries; many are incredibly abstract, and most have little or no relation to the real universe. Fortunately, the universe provides a few clues as to which symmetries are going to work.

Take the weak force. Please.

The weak force is absolutely vital to our existence. It's the engine that churns hydrogen into helium and, in the process, turns protons into neutrons. Those are the particles that typically get the most attention, but the minor players deserve attention as well: positrons and neutrinos. It is telling that neutrinos or antineutrinos *always* seem to be involved whenever there's a weak interaction. They seem to always show up in places where an electron might also seem at home.

Neutrinos are *very* closely associated with electrons. We get a pretty good hint of this in the particle zoo. The fermions are collected into pairs. This is not just a convention; it's another symmetry.

The mathematicians have a special name for the Electron–Neutrino Symmetry. They call it SU(2). It might comfort you to know that we've actually seen this symmetry before, but in a very different context. It's exactly the same symmetry that describes spin. Electrons can be spin-up or spin-down or some combination of the two. We've also seen that it doesn't matter which state an electron is in. If I turned all ups to downs and vice versa, all of our interactions would remain essentially the same.

The similarity is so perfect that the equivalent of electric charge is known as weak isospin. Just as a spin-up and a spin-down electron both have a total spin of ½, regardless of the direction, *up* in this case corresponds to a neutrino, and *down* corresponds to an electron, and the weak force is capable of converting the one flavor to the other. If you turned all electrons to neutrinos and vice versa in the entire universe, the weak force couldn't care less.

This is a very strange revelation. Electrons and neutrinos normally

---

* And yes, I realize that it already sounds very hard.

look very different. That's because most of our world is dominated by electromagnetism, which is enormously stronger than the weak force. In electromagnetism, the electron and neutrino are *very* different. One has charge, and the other doesn't.

The point is that we have a symmetry, and from this symmetry, we get a conserved quantity:*

Electron–Neutrino Symmetry → Conservation of weak isospin

The weak force behaves almost exactly the same way as electric charge does in electromagnetism. It tells us how various particles interact with one another. Also, because the weak force is a little more complicated, there's also *another* property called weak hypercharge, which looks a *lot* like ordinary electric charge if you don't look too closely.

We also get mediating particles. In the weak force, they're known as the W^+, W^-, and Z^0 bosons, and as we'll see shortly, they do *not* behave as simply as we might have hoped. For one thing, there is an obesity epidemic among the weak force particles, one that Yang and Mills were not prepared for.

WHY DON'T ATOMS EXPLODE?

Before we clean up the Standard Model, we need to finish taking inventory. Electrons and neutrinos aren't the end of the story. You can't make an atom out of them, for instance. The constituents of atoms—protons and neutrons—exhibit a fairly obvious symmetry. As Reed College physicist David Griffiths has put it:

> There's an extraordinary thing about the neutron, which Heisenberg observed shortly after its discovery in 1932: apart from the

* Thanks, Emmy Noether!

> obvious fact that it carries no charge, it is almost identical to the proton. . . . Heisenberg proposed that we regard them as two "states" of a single particle, the nucleon.

Neutrons and protons differ in mass by only about 0.1 percent. They also, as we've seen, have a very intimate relationship in that the former can decay into the latter. In a way, this isn't much of a surprise, because protons and neutrons are made of very nearly the same recipe.

There are, as you already know, a bunch of particles known as quarks. They were first discovered experimentally in 1967 at the Stanford Linear Accelerator (SLAC), though they'd been suspected for some time before then. The two types that most concern us are the up and down variety: two ups and a down for a proton, and two downs and an up for a neutron. In other words, at the end of the day, turning a proton into a neutron is really the same thing as turning an up into a down.

The relationship between the up and the down quark seems very much the same as the relationship between the electron and the neutrino. But in another way, quarks are very *different* from electrons and neutrinos.

QUARK CONTENTS

PROTON

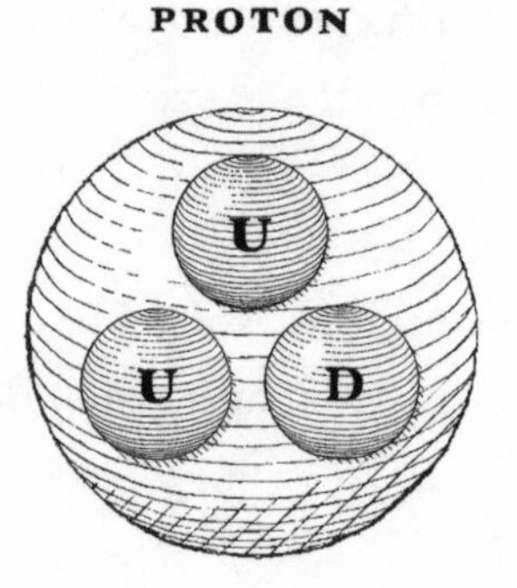

NEUTRON

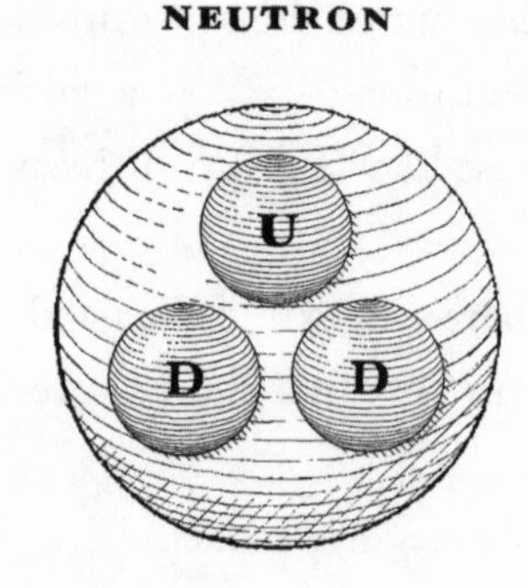

Just as we split up our Standard Model into fermions and bosons, we're going to split up our fermions into two piles as well. Take a peek at the particle table in the back of the book, and you'll see I've already done so. In one pile we have the electrons, muons, tau particles, and neutrinos—collectively known as leptons—and in the other we have quarks. The big difference between the two is quarks are subject to the strong force while the leptons are not.

If you're wondering why it is that there are two incredibly similar piles of particles, one with the strong force and one without, I honestly can't tell you. While we've managed to bundle up electromagnetism and the weak force together, we haven't entirely figured out how the strong force gets bundled in.

The strong force holds the nucleus together. This is no small feat. After all, even a single proton is a powder keg waiting to go off. Two up quarks each have a positive charge, which means that according to electromagnetism, they should repel each other. The strength of repulsion is gargantuan—roughly 10^{30} times stronger than earth normal gravitational acceleration—which means that you need something even stronger to hold the whole thing together.

As with the weak force, the symmetry of the strong force has a special name: SU(3). And as with the weak force, the symmetry predicts a conserved quantity: color. This is basically the equivalent of charge, but for quarks. The weird thing is that *any* particular type of quark can have one of three possible colors: red, green, or blue. Just to be clear, this is only a name. If you could zoom in at subatomic scales, a blue quark would look the same as a red one. We could have just as easily picked any other group of three: lawful, neutral, and chaotic, for instance.

The idea of three different charges is alien to people accustomed to dealing only with plus and minus, but you are forbidden from freaking out about this. Colors are exactly like electric charges, except that there are two opposite types of charge and one same charge.

It's no coincidence that there are three colors; it's just a consequence of the *3* in SU(3). Leptons have no color, so in the same way that electrically neutral particles are ignored by electromagnetism, leptons sit out the strong force.

This symmetry says that if you changed red particles to green and green to red (or any other switch), then the interactions would be the same as before you started. Or to put it another way:

Color Symmetry → Conservation of color

Yeah, I know. The thing that's conserved is essentially the same as the thing that's symmetric.

One of the strangest things about the strong force is that all the naturally occurring particles in the universe seem to be colorless. A proton will have a red, a green, and a blue quark. If you've ever played around with color addition, you'll note that adding up all of the colors gives you white, no color at all. This is why protons and neutrons need to have three quarks and not some other number.

Just as with all of our other Gauge Symmetries, color leads inexorably to mediator particles known as gluons. At first blush, gluons play roughly the same role as photons do in electromagnetism. When two charged particles want to attract or repel one another, they send photons back and forth. Likewise, two quarks will pass their notes via gluon.

There is an important difference, however. Photons are themselves neutral, which means that the two photons won't interact with one another. Gluons are not so fortunate. Have you ever seen a small child try to pull a small strip of tape off of a roll? If the answer is yes, you'll have noticed that the entire thing ends up in a tangled mess of bunches and blobs. Gluons directly interact with one another, and consequently tend to get in each other's way. This, incidentally, is why the strong force is confined to the nucleus of atoms.

WHAT'S SO SYMMETRIC?

As elegant as the Standard Model may be, there is still a lot to keep track of. Perhaps a little table of the symmetries of the Standard Model is in order:

Symmetry (According to the Mathematicians)	Interaction	Conserved Quantity	Mediators
U(1)	Electromagnetism	Charge	Photon
$SU(2)_L$	Weak	Weak isospin Weak hypercharge	W^+, W^-, Z^0
SU(3)	Strong	Color	Gluon

You may have noticed that I slipped a weird little *L* into the symmetry for the weak force. What's that, you ask? It's a remnant of something we saw earlier: Neutrinos are *always* left-handed. Left-handedness is *intimately* related to the weak force. Right-handed particles are totally immune to the weak force (just as colorless particles are immune to the strong force, and neutral particles are immune to electromagnetism), which means that in a very real sense, left-handed and right-handed particles of the same type really are distinct creatures. This little asymmetry is going to be very important in short order.

All of this seems a little list-y and a long way from what you might normally think of as symmetric, so it might help if we draw all of the Standard Model particles on a single diagram. We've done so on the next page.

Beautiful, right?

This is just one way of drawing the various charges of all of the particles. In this particular case, every point on the diagram corresponds to a particular combination of weak isospin and weak hypercharge. If you know how to combine the two, you also get ordinary electrical charge out for free.

It doesn't take long before you notice that the particles themselves fall into a very regular pattern. If we had been lazy in detecting particles and had somehow missed a few, the holes in the diagram would tell us pretty quickly where to look and what sort of properties those particles will have. These diagrams are also pretty helpful because we can immediately see all of the conserved quantity within a particular law.

But because I can draw pictures only on a two-dimensional page,

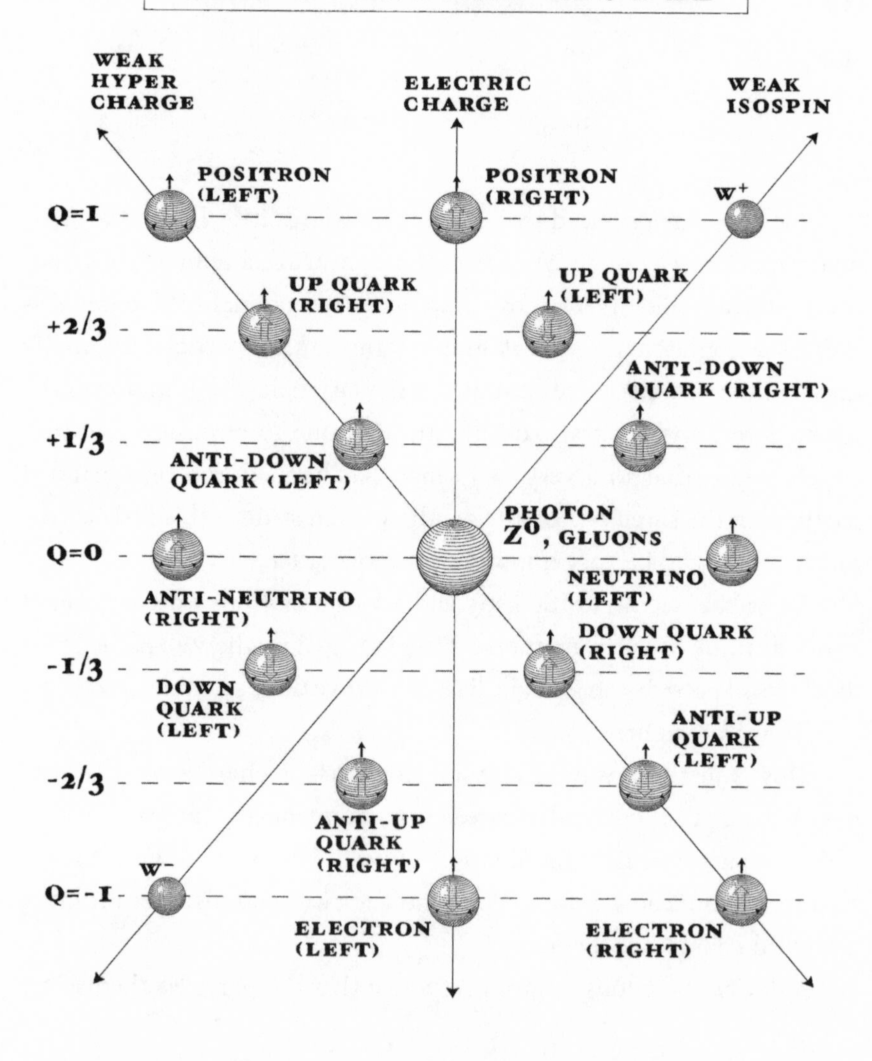

there is a lot of information missing from this diagram. For example, quarks can have any of three colors, and yet a red quark will be at exactly the same place on our diagram as a blue one. In other words, depending on how we look at particles, we'll see different symmetric arrangements.

THE ELECTROWEAK FORCE AND BEYOND

These symmetries are more than just math tricks. In 1960, Sheldon Glashow recognized that the weak and electromagnetic force could be combined into a single "electroweak force," a theory that was perfected later in the decade by Abdus Salam and Steven Weinberg. This is one of the nice things about symmetry. Take a look at the relation between the weak isospin, the weak hypercharge, and ordinary electric charge, and you'll note that they are *very* closely related. This is a pretty good clue that the underlying forces are probably unified too.

Unification is a big deal. For one thing, it's a real time-saver. If all of the laws of physics can be described (ideally) by a single equation, you don't have much to memorize. It also means there's some serious internal consistency to physical laws.

Newton is remembered centuries later because he was able to unify the motions of the planets, the swinging of pendulums, and the falling of apples into a single law of gravity. Likewise, electricity and magnetism look very different on the surface. Electricity governs the interactions of balloons on sweaters, whereas magnetism controls compasses. But when you, provided you are Maxwell, look deeper, the distinction between the two is just a matter of deciding whether the charges are moving.

Electroweak unification is a bit trickier, but the upshot is that at early times, there was a single force described by a single equation, but with *four* mediator particles. It was only the cooling of the universe and the rather mysterious Higgs mechanism that makes the two forces appear separate.

In the unified Electroweak Model, the photon and Z^0 aren't really two different particles but are two different states of the same particle.

Why shouldn't they be? Both are electrically neutral. Both are spin-1. And while the Z^0 has considerable mass today, in early times, both were massless.

In other words, the photon and Z^0 looked the same and interacted with particles in much the same way. Rather than the charge we all know and love, they responded to a particle's hypercharge (a combination of what we now call weak hypercharge and ordinary electric charge).

Once the universe cooled sufficiently, the photon and Z^0 looked very different from one another. In the process of breaking electroweak into "electro" and "weak," certain particles will end up interacting more with one than with the other. After the breakup, the neutrino, for instance, found itself responding to *only* the weak part and (being neutral) completely blind to the electro part.

This isn't as strange as you might think at first. There are lots of ways that you can divide up a pile of particles, and your method of sorting will depend on what you want to do with them. A pile of coins could be grouped by which side is up, but it's pretty clear that it makes a lot more sense to group them by denomination. Because the energies involved with photons and Z^0s are so different from one another, today they are grouped as two distinct particles.

This is what happened when the electroweak force broke up, but here's *why* they broke up. It's all about mass. One of the predictions of the Yang–Mills theories is that all mediator particles should be massless. The photons and gluons have absolutely no problem abiding by this rule. Unfortunately, Yang and Mills seem to batting only two for three.

The particles of the weak force are the heavyweights of the mediator world.

To put things in perspective, W bosons are roughly 85 times the mass of a proton, and the Z^0 is even more massive than that. The weak mediators are supposed to be completely massless, and instead they are *huge*. This hugeness plays a big role in the physics of the weak force. It is, in fact, what makes the weak force so weak. As Glashow put it:

It is a stumbling block we must overlook.

We can't entirely overlook the masses of the Ws and Z^0. These huge masses are generated by the Higgs boson, a particle that's gotten a lot of attention lately. The Higgs is the Yoko Ono of the electroweak force; it was the catalyst that split the electro and weak in two. But to really understand why the Higgs is such a big deal, we first need to say a few words about where mass comes from.

MASSES AND FIELDS

Einstein is given a lot of credit, and rightly so. He showed that time is relative, that light is a particle, and that the atom is actually a thing. But he's most famous for his equation showing that you can get energy out of mass, and, even more important for our purposes, that you can get mass out of energy:

$$m = E/c^2$$

I've described fields as ripples in various cosmic trampolines, but I haven't given very much attention to how one trampoline differs from another. And they *do* differ. Some trampolines are stiff, whereas others are a bit more yielding. A stiff trampoline is tough to get going, but once you do, you get quite a bounce out of it. That sounds almost exactly like mass! It takes a lot of effort to get a massive particle moving, but once you do, it takes a lot of effort to stop it.

A trampoline is a simple two-dimensional model of the universe. If you want to make things even simpler, and step down to one dimension, you can do so pretty easily on a guitar. Very thin strings are also very light, and extremely easy to pluck. They oscillate very fast, and produce a high-pitched tone. The thick ones are more unyielding, and produce a lower tone. And each string corresponds to a different

particle. Pluck them at different intervals or with different amounts of force, and you can get some very complicated sounds.

To figure out how all of this works in the realm of fields, we have to zoom in and look at how a small patch of a trampoline behaves under stress. Whether a trampoline or a rubber band or a guitar string, there's a general tendency for elastic materials to want to restore themselves.

Consider the myth of Sisyphus. Sisyphus, as you may recall, was the king of Corinth and an all-around dirtbag. After taunting the gods once too often, he was sentenced to push the same giant boulder up a hill for all eternity.

Suppose you (as Sisyphus) want to start a particular field ajostling. This is a necessary prerequisite for creating particles. On level ground, it's very easy to jostle a boulder from side to side. In a steep valley, on the other hand, it's much more difficult.

Sisyphus and the trampoline are two different ways of looking at the same thing; it's just that the trampoline gives you the global view of a field, whereas Sisyphus sees things more locally. It doesn't matter whether he decides to push to the east or west; it's equally difficult either way. This too is a symmetry, but a fairly abstract one. Sisyphus isn't moving his boulder through space. Rather, pushing to the east is the equivalent of plucking the trampoline in an upward direction, while west is down.

The shape of the valley is entirely determined by what kind of particle you've got, and the steeper the valley, the more massive the particle feels, essentially because it becomes more and more difficult to jostle the field.

Make sense? Good. Because I'm about to blow your mind.

Not every field is so simply described by a nice, smooth valley. Some are complicated mountain ranges of peaks and valleys. Unless you put energy in, a boulder will eventually come to rest in a valley, but you never know whether you're really at the bottom or if you're just trapped in a ditch.

But from the perspective of Sisyphus, it doesn't matter whether you're really at the bottom of a mountain or just in a small valley. All that matters is how much effort it takes to budge the boulder.

Put another way, the apparent mass of a particle can *change.* It all depends on the shape of your mountain and where on the mountain you

MASS AND POTENTIAL

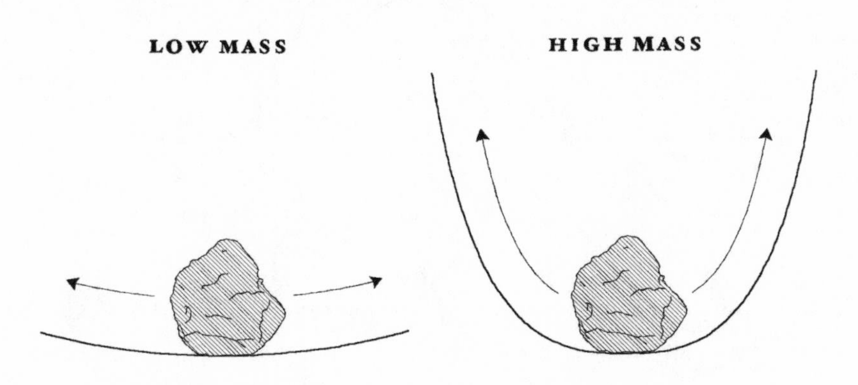

happen to find yourself. The complex mountainous terrain of these fields will turn out to be the key to unraveling the mystery of why the weak force is so weak.

SYMMETRY BREAKING

Let's say you (still as Sisyphus) decide to climb an unusually symmetrically conical volcano. You ascend to the top, precariously balance your boulder at the peak, and promptly fall asleep.*

When you awake the next morning, you're disoriented. You don't have a compass or GPS, clouds blot out the sun, and mists obscure the world below. It is a very bad start to the day.

Your boulder is in a very unstable situation. The slightest nudge in some random direction will send it down the side of the mountain.

HIGGS MOUNTAIN

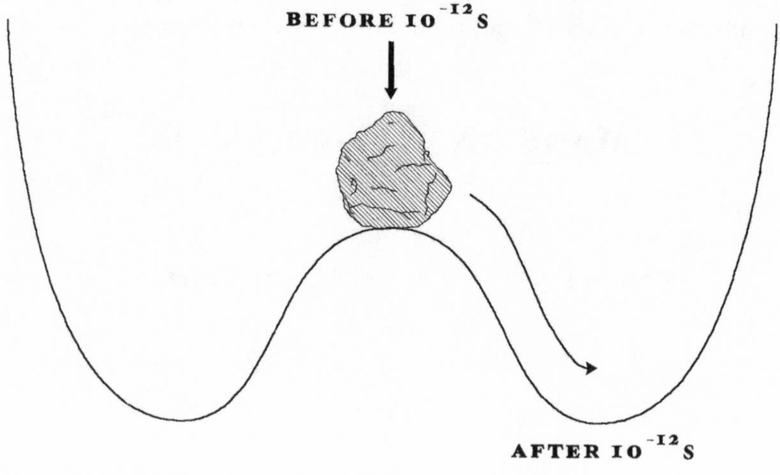

BEFORE 10^{-12} s

AFTER 10^{-12} s

* This probably isn't a great idea. It seems more like something that Wile E. Coyote might do.

Because nothing could be easier than nudging the rock and start it rolling, the corresponding particle in this analogy must be massless.

The boulder rolls down the mountain and into a valley. Getting it out of the valley is an incredible chore. The mountain, of course, hasn't changed. It's still perfectly symmetric. But because you're no longer standing on the peak, the mountain doesn't *look* symmetric. What's more, boulders in valleys correspond to massive fields. Just by falling down the mountain, a particle went from massless to massive in one fell swoop.

This is an example of *symmetry breaking*, and it is *the* key to understanding why the laws of physics today aren't the same as they were in the beginning of the universe.

At very high energies, the kind that filled the universe in the first 10^{-12} second after the Big Bang, electromagnetism and the weak force were unified, as we've seen. The Ws, Z^0, and photons were all mediators of the electroweak force. Meanwhile, there was another field hanging around: the Higgs. It is the Higgs field that started off on top of the mountain.

The Higgs boson has gotten a lot of press over the last few years. We (the physics community) made a huge deal over it because it is the *last* particle predicted by the Standard Model *and we found it*! The Higgs is the answer to a riddle of why the weak force is so weak.

It's also important because it explains the existence of masses of the Ws and Z^0 bosons. Massive mediators can't travel very far, which is why two hydrogens need to get *very* close to one another before they can even *think* about fusing (it's the weak force at work, remember). A molecule of water (two hydrogens and an oxygen), for instance, would have to be squeezed by roughly a factor of a million before the weak force kicks in. Good thing too. You're 70 percent water, and it would be a real shame if you spontaneously underwent nuclear fusion.

In 1964, half a dozen researchers—Robert Brout, François Englert, Carl Hagen, Gerry Guralnik, Peter Higgs, and Tom Kibble—realized that symmetry breaking may hold the key to the mass of the W and Z particles.* The idea seemed at first to be kind of absurd, even to its originators. As Higgs himself wrote to a colleague:

> This summer I have discovered something totally useless.

The Higgs mechanism tells a story. It begins, as we've seen, on top of a seemingly symmetric mountain. The Higgs particle sat precariously on the top, but just as neutrino and electron are two sides of the same coin, the Higgs has—or at least, had—this dual identity as well.

This is important to keep in mind. Remember that the Electron–Neutrino Symmetry is directly related to the weak force. Whatever else the Higgs particle does, it will also interact with the W and Z particles.

The universe in the very early days was so hot that it was as if a huge number of people were vigorously jumping on the Higgs trampoline or, equivalently, as if Sisyphus were pushing his boulder around in random directions like a maniac.

It was too hot, in other words, for the Higgs boulder to roll very far down the mountain before it got a push back up. As the universe cooled, Sisyphus settled down a bit, and the boulder started to roll down, eventually falling into a relatively deep valley. This descent breaks the beautiful symmetry that we started with. While initially the Higgs could have gone in any direction, rolling down the mountain essentially picks one and locks that direction in as special.

The Higgs field forgets entirely about being at the top of a mountain, and it quickly settles into life at the bottom of a valley. But fields trapped in a valley, as we've seen, mean that the corresponding particle has a mass. Just by symmetry breaking, the Higgs gains mass out of thin air. This *is* a surprise.

* Higgs generally gets credit (name-wise), though he has advocated simply calling it the H-boson.

HOW THE HIGGS MAKES MASS

Of course the Higgs mass was really created from pure energy. That was Einstein's gift to all of this. But the Higgs isn't famous because it *has* mass; it's famous because it *gives* mass to other particles.

If you want the quick-and-dirty version of how the Higgs creates mass, then just remember that the Higgs field interacts with other fields. Interactions, in the realm of physics, mean energy, and energy means mass. In the language of trampolines or its simpler cousin, the guitar string, you can think of the Higgs as you would barring the frets; it raises all of the frequencies and, consequently, all of the energies. That extra energy is what we think of as particle masses.

But that description is a little too pat. To really understand how the Higgs creates mass, we need to figure out how the Higgs field changed over time. Various fields interact with one another. The electron field interacts with the photon field. This is, of course, the electromagnetic force. Likewise, the Higgs is tethered to the other fields. Normally, this wouldn't be such a big deal. In empty regions of space, we'd expect the interaction with the Z^0 (for instance) to completely vanish because, on average, the Higgs field will completely vanish.

But once the Higgs rolls down the mountain and into the valley, all bets are off.

Remember that the Higgs field, in particular, responds to the weak force. Once the boulder rolls down the mountain, there's a constant field and thus a constant interaction energy between the Higgs and the W and Z^0 fields. But these interactions work both ways. Remember:

> **Newton's Third Law:** To every action there is always an equal and opposite reaction: or the forces of two bodies on each other are always equal and are directed in opposite directions.

Just by adding a constant interaction—*pow!*—we get a mass. You can think of this as an example of the adage A rising tide raises all boats. Because the Higgs is no longer at the top of the mountain, none of the fields that it couples to is in *its* equilibrium condition either. The Higgs and the weak mediators are like the Wonder Twins of the subatomic world. Once one of them changes properties, they all do.

There's a useful way of thinking about all of this. The Higgs is a boson, which means that it gets grouped with the photon, the gluons, and the other mediators—the middlemen of the particle physics world. While most physicists wouldn't describe the Higgs as a mediator, you can think of it as one—just one that allows a particle to interact with *itself.* A W^+ particle continuously sends and receives Higgses, except that instead of sending the Higgs particles somewhere else, it reaps what it sows, in the process working up a sweat of interaction energy. Since energy is mass, the W^+ gains weight, seemingly from nowhere.

Put into a bit of mathematical detail, the Higgs mechanism makes a number of predictions, most notably that the W and Z^0 particles should have mass. The prediction is deeper than that, though. It also suggests the Z^0s are going to be 1.14 times that of the Ws, exactly what is found by experiment.

I realize that the whole picture of coupling fields to one another is not exactly the most intuitive concept in the world. There has been a cottage industry built around the idea of trying to describe the Higgs in terms of simple metaphors.

For instance, some scientists will liken the Higgs field to a cosmic pool of molasses. As particles move through it, they acquire a resistance to their motion, just like mass!

At first blush, this seems like a great picture, until you realize that it leaves a lot of questions unanswered. For instance, why should it be that only some particles interact with the Higgs molasses, and why don't they interact equally? You and I would be equally impeded by swimming through a thick, syrupy pool, but a Z^0 is far, far more massive than an electron. And a photon, of course, is completely massless.

The analogy breaks down once you see how it would play out in the real world. Try swimming through molasses. You'll continually slow down until you come to a stop, and you know that's not how particles really move. If you remember one thing from your high school physics class, it's probably Newton's old (and still true) *Objects in motion stay in motion*.

There are loads of other analogies. One popular one involves a celebrity walking into a party. When she walks in, she's immediately swarmed by admirers, greatly impeding her progress and significantly increasing her mass. You, on the other hand, can proceed unimpeded. You are meant to be a photon, while the celebrity might be a Z^0.

The Higgs (the fans) couples to the celebrity, but not to you. Once in motion, the celebrity is then pushed along slowly by her admirers, making stopping difficult. The problem is that the Higgs is one of the heaviest particles, which would mean that there was no explanation for *any* massive particle to be lighter than the Higgs.

Peter Higgs has likened his eponymous field to the index of refraction of glass. Simply by moving through glass, light goes slower than *c*. Because light travels at a constant (albeit slowed) speed through glass, water, or another medium, at least Newton's First Law seems to be satisfied. The problem is that with enough effort you *can* mostly overcome the limits of mass. Particles have mass, but we're still able to accelerate them to speeds greater than 99 percent the speed of light.

The point is that no matter the analogy, you're likely to encounter a problem. Your gut intuition makes it seem as though mass is fundamental. The idea that mass can be imbued through interaction energy just seems anathema to our everyday experience. It's okay. We'll see in short order that *most* of the mass in the universe, if not all of it, is really nothing more than an illusion.

Of course, this entire house of cards is built on the notion that the Higgs actually exists.

ARE WE QUITE CERTAIN THERE *IS* A HIGGS?

Predicting the Higgs is one thing, actually detecting the Higgs is quite another. In 2010, operations began at the Large Hadron Collider (LHC), an accelerator in Switzerland and France, 27 kilometers in circumference that takes protons up to 99.999996 percent the speed of light and smashes them into one another.* Just to put those numbers in perspective, the time dilation at those speeds are so high that the internal clocks of the protons measure only about 1 second for every *hour*, according to the scientists.

The LHC is one of the largest scientific collaborations in the history of science, with literally thousands of physicists involved in the project and a start-up cost of roughly $4 billion.

To be sure, finding the Higgs wasn't the only motivation for the LHC. There remains the hope that we might discover the elusive dark matter particle or find a supersymmetry underlying the Standard Model of physics. There are all sorts of high-energy phenomena that we could never previously test. But the grand prize was always the Higgs.

Protons going as close to the speed of light as those in the LHC tend to do a lot of damage when they strike one another. In the resulting maelstrom, lots of high-energy particles are created, including, as it happens, the Higgs. The Higgs itself is incredibly difficult to detect. It's neutral, after all, which means that it doesn't give off any radiation. This may be something of a surprise. When the newspapers announce that the LHC or some other experiment has detected a particle, it often means something very different from what you might suppose. At no point were the scientists on the experiment able to put a Higgs in a Petri

* It's all done with magnets.

dish or even see its trajectory in a bubble chamber. Instead, the Higgs is detected by noticing that two high-energy γ-rays* seem to have emanated from a single point, and by using the conservation of momentum and energy, the mass and motion of the particle were computed.

In July 2012, spokespeople from two experiments, A Toroidal LHC Apparatus (ATLAS) and Compact Muon Solenoid (CMS) announced the discovery of the Higgs, one of the most important particle discoveries in the last 50 years. They found that the Higgs was about 133 times the mass of the proton.†

Now, my lawyers insist that I tell you that the new particle seems to have all of the markings of a Higgs, but there remains much to do before we can say unambiguously that it *is* the Higgs. We know, for instance, that it's a boson with either spin-0 (Higgs!) or possibly spin-2, which can't be ruled out. We know that it has a mass comparable to the Ws and Z^0, which isn't a surprise, considering what the Higgs actually does. The upshot is that, while we can't be completely complacent, most physicists take it as given that what we're seeing is the Higgs.

Upon its discovery, the popular press went nuts. The *New York Times* headline read, "Physicists Find Elusive Particle Seen as Key to Universe." Almost every newspaper and magazine article referred to the Higgs as the God Particle.‡ They described it in awed tones that suggested that we were nearly at the end of physics and that we now understood the nature of matter. Bring your mint-condition *Millenium Falcon* in its original packaging to Comic-Con, and you'll get a similar reaction.

---

* Mostly. The so-called gamma-gamma channel was the strongest signal, but occasionally the Higgs decays into two Z^0 particles as well, and they subsequently decay into two pairs of electrons and positrons or muons and antimuons.

† If you read semitechnical discussions of the Higgs on the Internet, the mass is usually given as 125 GeV. Because mass and energy are equivalent, this is just a fancy way of saying how much energy it would take to make a Higgs from scratch.

‡ Again, please don't do that.

And while discovering the Higgs is a *very* big deal, it turns out to describe surprisingly little of the mass in our everyday life.

WHICH MASS COMES FROM THE HIGGS AND WHICH DOESN'T

The Higgs mechanism was concocted to explain how it could be that the W and Z^0 bosons had mass even though none of the other mediators do. The explanation was that there was a spontaneous symmetry breaking in the early universe—a symmetry breaking that, among much else, made electromagnetism and the weak force separate from one another.

But this does *not* explain, at least not immediately, why it is that other particles have mass, and symmetry arguments would suggest that all of the fermions *should* be massless. It's a good thing they aren't. If the electron were massless, it would be impossible to make any stable atoms or molecules.

In the first chapter, we saw that our universe has a broken P Symmetry. Physics in the rearview mirror seems to look ever so different from our own.

The culprit, as you may recall, was the weak force. Whenever the weak force is involved in an interaction, any and all neutrinos that are produced fly out left-handed. In other words, they appear to be spinning clockwise as they head toward you. This left-handedness is true for *all* fermions in a weak reaction, or would be, if they were massless. They aren't, so these asymmetries aren't perfect. This definiteness of spin is intimately tied up to other symmetries we've seen before and, in particular, to the speed of light.

You can't outrun a massless particle, as it always travels at the speed of light. For instance, no matter how fast you run, your mass (however small it is) always prevents you from running from an incoming photon so quickly that the distance between you increases.

What distinguishes a left-handed particle from a right-handed particle is how it appears to spin as it heads toward you. With a massive particle, simply by changing my state of motion, I can make it appear as though an approaching particle is a receding one. In one fell swoop, and just from perspective, I can turn a left-handed particle into a right-handed one.

But if the weak force *really* obeys the left-handed symmetries that we've seen—a symmetry, I remind you, that is ultimately responsible for the dominance of matter over antimatter—then the only way for the symmetry to hold is for all participating particles to be massless.

The quark, the electron, the neutrino—all of them should be massless, but none of them is. The masses of the quarks, of the electrons, and most likely, of the neutrinos ultimately come from the Higgs field as well. Our understanding of the exact mechanism is still a bit murky, but now that the Higgs is established as real, we've got a pretty good sense that we're on the right track.

As important as it is, however, the Higgs doesn't "give rise to everything in the universe" as it is occasionally promoted. Most important, it doesn't *really* give rise to *your* mass.

You are made of protons and neutrons, and your protons and neutrons are made of quarks. But, as we've seen, the whole is much more than the sum of the parts. The total mass of quarks in a proton is only about 2 percent the mass of the proton itself. The rest, virtually all of your mass, is made up of the interaction energies between the quarks.

And that's just your atoms. The Higgs also doesn't tell us where Dark Matter comes from, that makes up 85 percent the mass of the universe.

And while the Standard Model itself has proven a marvel of precision and prediction, there's still something a little fishy about a fundamental theory that has nineteen entirely tunable numbers in it. We know these numbers only because we *measure* them.

My point is that discovering the Higgs may mean the end of the Standard Model, but it is most certainly not the end of physics. And while (spoiler alert!) we still haven't *reached* the end of physics, we have a sneaking suspicion that symmetry will help us get the rest of the way there.

Chapter 10

HIDDEN SYMMETRIES

IN WHICH OBJECTS IN THE MIRROR ARE CLOSER THAN THEY APPEAR

One of the great injustices in the world is that physics, and science in general, has a reputation for an unpleasant exercise in abstract navel gazing. This is informed, no doubt, by exercises involving pulleys and blocks on planes that students might see in their high school physics classes. It's as if we were to judge the pleasure of music based on nothing more than practicing scales.

Science is supposed to be fun. Oh sure, every now and again, we get bogged down in lists of particles and fairly arcane-sounding rules, but you should never for a moment lose sight of the fun. At its best, physics is a game that spans the entire universe.

Treating science as a game may seem to trivialize it, but I think the rules are what make a game worth playing, quidditch notwithstanding. Understanding the rules is the first step toward mastering the game.

In the physical world, we don't actually get the rules in a handy book. We have to infer them from observation and experiment. And as we've bumped against the limits of what we know, we're going to need to step back for a moment and take stock of what we might be missing.

Early on, we saw Richard Feynman's description of science as a game of cosmic chess. It's only by observing game after game that we can eventually discern all of the symmetries in the universe, and perhaps even more important, we can see when those symmetries are broken.

We can play, and have played, the same sorts of games with the laws of the universe. From just three internal symmetries we immediately get a list of all of the particles and forces in the universe. We can give you a list of *all* of the possible fermions (those are the spin-½ particles that make up matter, remember) that can possibly exist. We can predict all of the fundamental forces, and the bosons that serve as their mediators. We can figure out which particles have particular charges or colors or weak isospin or hypercharge.

Not bad.

Every single particle predicted by the Standard Model has actually been found, with no extras. Beyond that, we can calculate all sorts of interactions to insanely high ten-digit accuracy.

But for all that, to the untrained eye, the Standard Model still looks pretty *inelegant.* Some of these rules are awfully ad hoc and, frankly, leave most people who've thought about them with the unsettling feeling that there might be something much, much deeper. The Standard Model is a cool apartment where you've filled in all the holes with toothpaste in the hopes of getting your security deposit back. Oh sure, it looks great, and for the most part it *is* great, but there are definitely some important bits that require an explanation.

MO PHYSICS, MO PROBLEMS

The Standard Model has its problems, as you'll see, but before I get into them, I want to make a little plea. There's a tendency to take this sort of confessional too seriously, as though every problem means that we need to start at square one. We don't. Whatever the flaws in our current model of the universe, we can't be *that* far off track. General Relativity is *more* correct than Newtonian gravity, but that doesn't mean that in a post-Einstein world, we walk around mocking Newton. In the same way, no matter what ends up unifying quantum mechanics and gravity, both theories will continue to produce outstandingly good predictions under normal circumstances.

That said, let me just give you a sampling from our buffet of ignorance, after which I'll give you a tour of our current best guesses for cleaning up this mess.

WHY SOME SYMMETRIES BUT NOT OTHERS?

Our entire model of the universe is built on symmetries. Some, like isotropy (the laws are the same in all directions), homogeneity (same in all places), and time invariance (same at all times) seem natural enough. Even relativity, the Lorentz Invariance that allows everyone to observe a constant speed of light, has an elegance to it that makes it seem natural.

But then we get into the internal symmetries of the Standard Model, and honestly, even to the initiated, they seem completely ad hoc (a.k.a. *ugly*). To be sure, Phase Symmetry (the one that gave us electromagnetism) is about as simple as you can get. But as for the others, the ones that give rise to the weak and strong forces, there are plenty of symmetries out there that are just as simple. Why'd the universe pick these?

There are symmetries that go beyond the ones that generate the forces. Look at the laundry list of the Standard Model again, and you'll note that all of the fermions are in three nice, orderly generations, each more massive than the last. The up and down quarks, for example, are the lightest. The charmed and strange quarks look almost identical—same charges, same spin, same interactions—but are of an order a hundred times more massive. The top and bottom quarks are the same, but *another* few hundred times more massive.

Why are there three generations of fermions, when virtually all of the interactions in our everyday life would have been perfectly happy with just one? As Nobel laureate Isidor Rabi put it upon the discovery of the muon—itself just a second-generation form of the electron:

Who ordered *that*?

It's not even as though the universe simply picked the simplest possible symmetries and went with those. For instance, Why is the universe left-handed?

Whenever a neutrino is created, it *always* spins the same way. This is more than a curiosity. It's incredibly important—to matter-based entities, at least—that the universe made some sort of choice, no matter how arbitrary. Andrei Sakharov showed us back in the first chapter that you can't have an excess of what the locals call matter in an ambidextrous universe.

Why and how did the universe choose one hand over the other one? There's no particular reason why we couldn't have a perfectly pleasant universe that's identical to ours, but right-handed rather than left-handed. And why is the weak force the only sinister one?

This isn't just a rhetorical question. One of the big lessons that we've learned from symmetries is that by and large if a theory *can* include some particular effect, then it probably will. In particular, the theory for the strong force quantum chromodynamics,* very naturally includes a term that violates reflection symmetry.

Think of it this way. Suppose you sit down at a round table at a fancy dinner party and you notice two water glasses, one to your right and one to your left. I'm sure Emily Post has some instructions

* The *chromo* just refers to the color of the quarks and gluons that shows up in the strong force.

on which glass you *should* drink from,* but supposing you're as ill-mannered as I am, either glass will do. But here's the thing: Once you pick up (say) the glass to your left, the guest to your left has no choice but to drink from the glass to *her* left and so on. Once the symmetry is broken, it's broken.

There is absolutely no evidence that there is *any* handedness in the strong force, and experiments have been conducted that would measure the asymmetry down to one-billionth.

An interesting potential solution was suggested by Roberto Peccei and Helen Quinn in 1977. They proposed that the left-right symmetry itself, as with other symmetries, gives rise to a particle. In this case, the particle is known as an axion. The axion would do more than just explain away the symmetry of the strong force. Because it's neutral, massive, and potentially quite numerous, the axion (if it exists) could be the missing Dark Matter particle. *Could* is the operative word. Despite exhaustive searches in both experimental and astrophysical settings, we've so far come up empty.

HOW STRONG ARE THE FORCES?

The names of the fundamental forces tell you a lot about their most important properties. Two of them are called weak and strong. We've seen why the weak force is so weak (the Higgs mechanism), but why is the strong so strong? If the strong force weren't stronger than electromagnetism, the quarks in your protons and neutrons would repel each other explosively, destroying you and everything you love. So while it's good for us that the strong force is strong, it doesn't really explain *why* it needs to be that way.

Even the weak force, despite its name, is actually about 80 percent *stronger* than electromagnetism. It appears weak only because the

* It's the one on the right.

mediators are so massive. Once you get those babies going close to the speed of light, the weak force looks a hell of a lot stronger.

And the relative strength of the forces is just the tip of the iceberg. All told, there are nineteen free parameters in the Standard Model. These include not only the strengths of the various forces, but also the masses of various particles and how forces and particles combine. None of these numbers ends up being simple mathematical numbers like 1 or π. Instead, they end up as unsightly numbers like 1/137.0359 . . . for the strength of electromagnetism, and 125 gigaelectron volts for the Higgs mass.

We've talked a bit about anthropic arguments for physical laws. These nineteen-plus parameters could be different values throughout the multiverse, and for some arcane reason that we haven't yet discovered, only a few of them (ours, for example) have exactly the right combination to produce complicated enough life to build particle accelerators.

It is also possible that we simply don't yet know enough about physics to predict the numbers from first principles and we just got lucky.

WHAT ARE THE MASSES (AND WHY ARE THEY SO SMALL)?

You may have noticed that I sneaked the masses of the fundamental particles into the list of unexplained parameters. "But wait!" you might exclaim. "Didn't you spend the entire last chapter *explaining* where mass came from?"

I did, but while the Higgs mechanism gives mass to various particles, the actual *amount* of mass needs to be put in by hand. One of the complications about discovering, and ultimately confirming, the existence of the Higgs is that even though we were fairly sure it was out there, there was no way of knowing exactly what its mass was. We just had to look at a bunch of different possibilities.

Stranger still, if you were to guess the masses of the particles, you'd almost certainly guess wrong.

Guessing particle masses isn't that different from what the guy at the fairgrounds does, except in our case, it's all about using the rest of the clues, the nature of our physical laws, to figure out how mass enters the equation. We can do this by mixing and matching all of the fundamental physical constants: c (the speed of light), $\hbar$ (Planck's constant), and G (Newton's constant of gravity).

These numbers are special because they're not particular to the strength of a particular force or a particular symmetry. Just as the speed of light could be set to 1 (making a light-year equal to a year), the other constants can be mixed and matched regardless of what units you pick. Multiply out the fundamental constants with the proper powers—and mathematically, there's only one way to do this—and you get a mass of about 20 billionths of a kilogram. This is known as the Planck mass,* and to a particle physicist, this mass is enormous, roughly 10^{19} times the mass of a proton.

Because it includes both the gravitational constant and Planck's constant, the Planck scale includes the effects of both strong gravity and quantum mechanics. In the *very* early universe, around 10^{-44} second after the Big Bang (an instant, incidentally, known as the Planck time), quantum fluctuations created black holes that literally spanned the entire universe. We *really* don't understand the laws of physics during the Planck time.

The Planck mass gives us a natural scale to expect for fundamental particles, but we've never discovered a particle that is even *remotely* close to the Planck mass. It's roughly 100 quadrillion times the mass of the top quark, the heaviest particle known. It's as if the guy at the carnival guessed that you weigh as much as Pluto. It's rude, sure, but it also suggests that he's probably in the wrong line of business.

If a proton, for example, has a mass of 10^{-19} that of the Planck mass,

* In case you're curious, the equation is $\sqrt{\hbar c/G}$, and it is quite literally the only way that you can combine the constants and get a mass.

physicists feel like the ratio is so minuscule that it requires an explanation. What are the odds that we ended up with something so tiny by mere chance? Because none of the known particles is even remotely the natural mass that they should be, the question remains, Why is everything so light?

HOW DOES GRAVITY WORK?

In describing our Standard Model, I've been using phrases like *the three forces excluding gravity*. But *why* exclude gravity? It's not as though it's unimportant in the grand scheme of things.

General Relativity does an outstanding job of describing gravity, but there's no denying that it has a completely different form—no mediating particle, no quantum uncertainty—from the other forces. How do we reconcile it with the others and, in particular, with quantum mechanics?

Because gravity dominates with large masses, and quantum mechanics dominates on small scales, the two theories don't typically butt heads. We've gotten a hint of how to incorporate quantum mechanics and gravity under normal circumstances through Hawking and Unruh radiation, but we still aren't sure how to unify the two theories in general.

We don't know how to deal with singularities like those found at the centers of black holes and at the instant of the Big Bang. The singularity is the cosmological equivalent of a bag of holding—you can pack literally infinite stuff into a finite space—and frankly nobody knows how that is supposed to work.

WHAT ELSE IS MISSING?

I made the rather bold claim that from the perspective of the Standard Model, we predict all of the particles ever observed, with no leftovers.

Technically, that's true, but I neglected to remind you that there are a couple of physical phenomena that still demand explanations, explanations that the Standard Model has been completely unable to help with. Unfortunately for us, these are not just minor corrections; they're Dark Matter and Dark Energy, which, combined, represent roughly 95 percent of the energy density of the universe.

Dark Matter, you'll recall, holds together clusters and galaxies, and seems to be five or six times as abundant as ordinary matter made of protons and neutrons. We are able to see its gravitational effects in a very direct way, leading to the natural conclusion that there should be some sort of Dark Matter particle flying about.

Because it makes up so much mass, there must be a *lot* of these Dark Matter particles. Dark Matter needs to be electrically neutral; otherwise it would be easy to see. The only Standard Model particles that could fit the bill are the neutrinos, and while they *are* quite abundant, they seem to be far too light to make up the Dark Matter. Axions, though not really part of the Standard Model, seem like they might be another option, but as I said earlier, we don't actually know that they are real.

There's an even worse problem, at least from the perspective of cataloging the energy contributions in the universe: Dark Energy, which seems to make up something like 73 percent of the total density. We can't just sweep it under the rug.

The simplest explanation for Dark Energy is that it is the net effect of particles popping into and out of the vacuum. In some ways, Dark Energy as vacuum energy is a perfect solution. Work through the equations, and it turns out that vacuum energy causes an accelerating expansion, just like Dark Energy.

But there's a problem. Of *course* there is.

The computed density of vacuum energy seems *way* too high. A straightforward calculation leads to the conclusion that it's about 10^{120} times larger than the *actual* Dark Energy density in the universe. In case you're wondering where this number comes from, the vacuum density is just 1 Planck mass per cubic Planck length.

Dark Energy is a worse problem than it first appears, because we don't even know what area of physics contains the solution. It could be that we're not interpreting the Standard Model quite right. Or it could be that Dark Energy is hardwired into the laws of gravity, Einstein's cosmological constant. If that's the case, we either have to accept Dark Energy as something that just *is* or else we're really not going to get a handle on it until we have a working theory of quantum gravity.

A reasonable conclusion is that we don't really have a handle on the dark universe at all. We can quantify it, which is certainly a good start, but we can't say much about what it really is.

SYMMETRY BREAKING, REDUX

Enough griping. We've gotten away from symmetry for just long enough to complain about all of the things that we don't know. But you shelled out good money for explanations, not excuses.

If we stretch our minds a bit, we realize that what's really going on here is that we have imperfect symmetries, like a perfectly imperfect Persian rug. Could it be that there was a time in the history of the universe when these symmetries were perfect and then something, the random chance of quantum mechanics, perhaps, simply tipped the balance?* Symmetry, in other words, broke.

Symmetry breaking has shown up a couple of times in our story already, but because we're thinking about the mind-bending world of internal symmetries and particle physics, a brief refresher might be in order.

Suppose you were observing Hoth, an ice planet. Wherever you travel on the planet, life is more or less the same, freezing cold. That's

* Answer: Probably.

because the planet is in the middle of space. The planet has a perfectly spherical symmetry. Life is the same no matter where you go, and while you're free to pick a North Pole or an equator if you absolutely insist on drawing a map, without a star or any other external landmark, those sorts of designations are more or less meaningless.

But if you throw Hoth into orbit around a sun, all of a sudden, everything changes. The equator, for instance, is a special place, and much like on the earth, will tend to be hotter than average. Climates will differ significantly by latitudes.

This sort of symmetry breaking has huge effects on human interactions on earth. In *Guns, Germs, and Steel,* the physiologist and geographer Jared Diamond argues that much of the flow of technology, agriculture, and disease followed lines of constant latitude, and that the east–west orientation of Eurasia conferred technological and immunological advantages on the inhabitants compared to the inhabitants of the Americas.

Simply by creating an interaction, we've gone from a 2-D symmetry, where the planet is much the same everywhere, to a 1-D symmetry, where life is the same only at similar latitudes. But unlike on Hoth, where the symmetries break by adding a heat source, it is almost always the case that *cooling* a system will cause a symmetry breaking.

Consider iron. You may know iron from its ability to hold your kids' artwork to the fridge. The spin of each iron atom forms a little magnet. While this is true for many materials, what makes iron special is that it is energetically favorable for atoms to align, and when they do so, those atoms add up to a very powerful magnetic field.

On the other hand, you can destroy an iron magnet simply by heating it to temperatures above 1,043K, the Curie temperature (after Pierre Curie). It's the thermodynamic equivalent of putting all of the iron atoms into a blender and totally randomizing the orientations. Starting with something that's decidedly asymmetric, a magnet has a north and south pole, symmetry can be restored by simply heating it up.

As the block of iron cools, provided it does so slowly enough, the

Cooling Iron

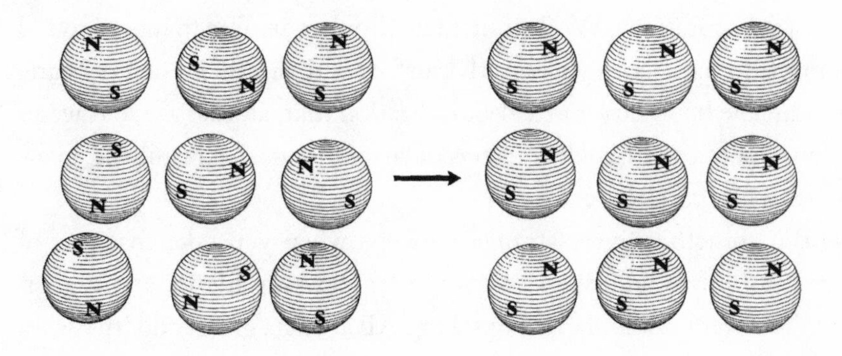

atoms will once again orient themselves parallel to one another, once again turning your lump of iron into a big magnet. Which direction will they settle in? Nobody knows. Of course, you could break the symmetry by hand by simply putting the iron in an external magnetic field, and everything will line up just as requested.

The same thing is (presumably) true of the laws of physics generally. The first 10^{-36} second of the age of the universe would have been a great time to be a physicist. Everything was so hot back then that the symmetries would have been obvious. Of course, it would have also been so hot that even our protons would have boiled away into individual quarks, but that's the price we pay for science.

LIFE AT LOW TEMPERATURE

This entire book has been about how symmetric the universe is supposed to be, but for all of that, the human world doesn't seem terribly symmetric. We don't live in an M. C. Escher drawing.

To take a simple down-to-earth example, our world has an up and a down and you can tell the difference between the two just by looking at how an apple falls or by collecting water in a pan.

A chemist could look at the interactions between water molecules and decide that no matter how you turn a droplet of water, the interactions should be the same. But that's not how it appears under normal terrestrial conditions. While individual droplets are free to move around from left to right as they please, the surface of the water presents a nearly unassailable boundary and a clear indication that, at least as far as water is concerned, the three dimensions of space are most certainly *not* symmetric with one another.

But something very strange happens when you take that pan of water and put it on a stove. Turn the temperature higher and higher, and the water boils and turns to steam. All at once, gravity doesn't seem to matter very much. The true symmetry of the interaction of the water molecules becomes apparent. The molecules start bouncing around in all three directions more or less equally.

The same thing is true with all of the fundamental forces of nature.

At very low temperatures—and *low* for this purpose can still mean hundreds of million degrees—the forces appear to be quite distinct from one another. But as you turn the temperatures higher and higher or, equivalently, as you turn the clock of the universe further and further back to the earliest moments, the underlying symmetries become manifest.

And what do those unifications include? The rather modestly named Electroweak Model describes the combination of electromagnetism and the weak force, but once we start adding other forces, we need to turn up our naming convention, rhetoric-wise. Grand Unified Theories describe the merger of weak, strong, and electromagnetism. Taking things a step further, Theories of Everything* add gravity to the mix as well.

Before getting into heavy-duty theories about complicated symmetries, it may help to step back for a moment and ask why, apart from

* Which, admittedly, sound like the sort of things you might think up in a lair hollowed out of the side of a mountain.

aesthetic arguments, we might suppose that the different forces are really the same.

Consider an electron all alone in the vacuum of space. Surrounding the electron, lots of particle–antiparticle pairs pop in and out of existence. These virtual pairs act like ripples on the ocean.

Though any *given* ripple will last for just an instant, at any given moment, there will be lots of them. During their short time in this world, virtual positrons will be attracted to the real electron, partially negating the electric field, while the virtual electrons will be repulsed.

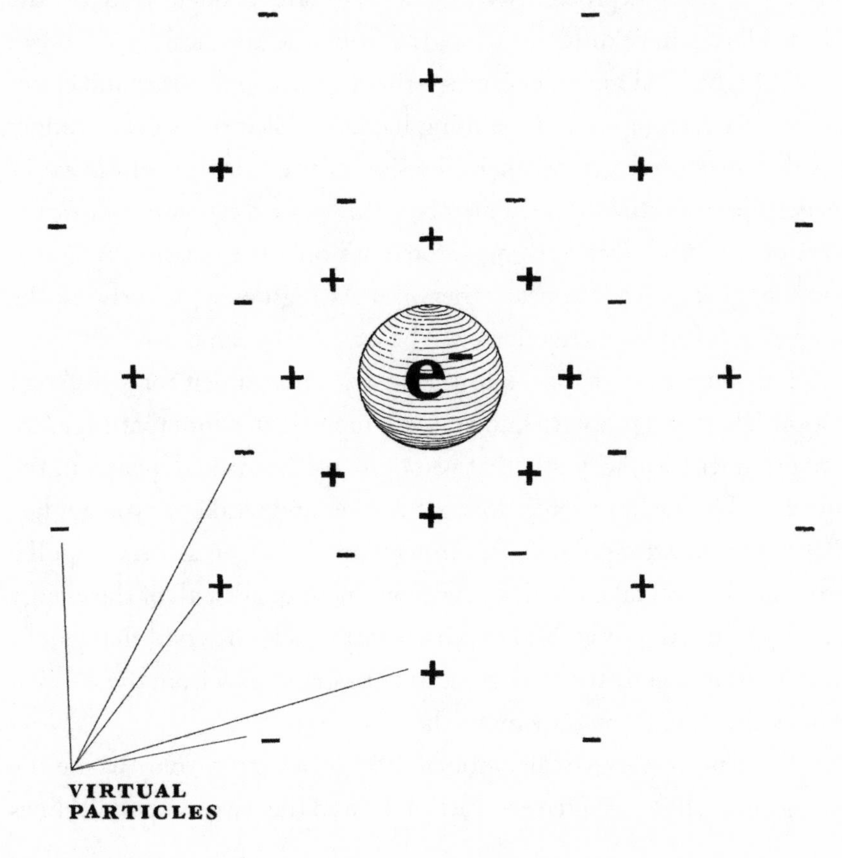

Think of this like sales tax. I know the dollar meal is supposed to cost a dollar and is advertised as such, but in Philadelphia, your Sausage McMuffin is really going to cost $1.07. This is the price you care about.

In the same sense, we don't really know (or maybe even care) what the bare charge of an electron is supposed to be, the number that we'd measure if we could somehow cancel out all of the virtual particles. Particle shielding acts more like a coupon than a tax (reducing rather than adding), but the effect is the same. The electron charge that we know and love and can look up in a book isn't the true list price. The electric charge you see is less than the bare charge, the one that we might see if we were somehow able to get arbitrarily close to an electron.

This discrepancy between apparent and bare helps shed some light on one of the big problems with the Standard Model: Why do the various forces have different strengths from one another?

At higher and higher energies—that is, if we look closer and closer to the bare charge—a strange thing happens. Electricity gets stronger but the weak force gets weaker. The sign of the shielding effect works in the opposite direction. Remember, the weak force is paradoxically *stronger* than the electromagnetic force; it's only the massive W and Z particles that make it appear otherwise. At high enough energies, the strength of the two interactions will be exactly the same.

The same thing happens to the strong force, albeit for a different reason. There is a property known as asymptotic freedom that explains, among much else, why it is that you don't see individual quarks in the universe. Unlike most forces that get weaker and weaker as you get farther away, the strong force gets stronger and stronger. If I try to pull a proton apart and take a look at the constituent quarks, all of the energy that I put in will just go into making *new* particles. It's odd that gluons seem to have exactly the same force as Sebastian Shaw from the *X-Men*. Attack them, and you only make them stronger.

There is an energy scale, around 10^{15} gigaelectron volts, where the strengths of all forces intersect. Euclid showed that two nonparallel lines

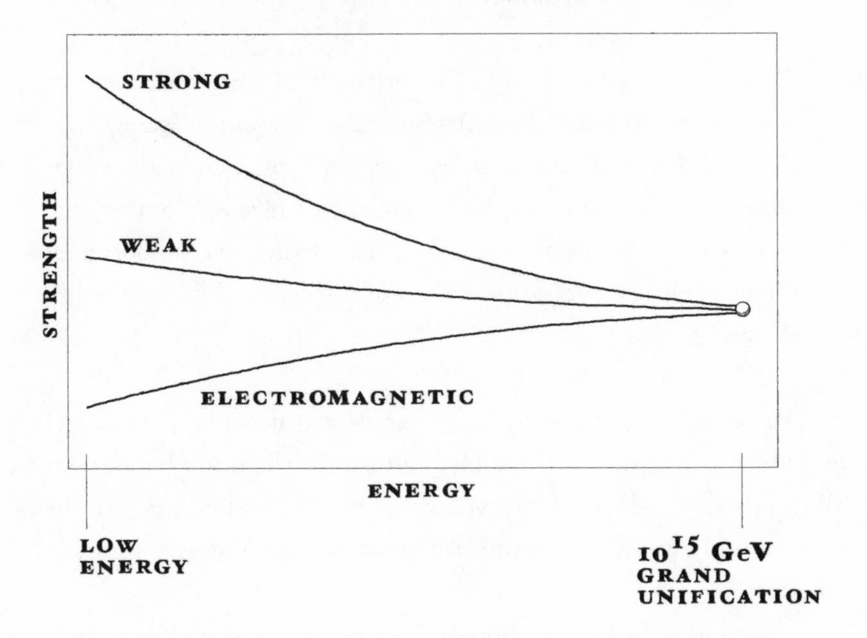

will always intersect at a point. But three lines should just intersect pairwise. It's strange, and a bit telling, that all should (as near as we can tell) intersect at a point.

Unfortunately for us, this point corresponds to an energy well outside of our technological abilities. We would need about a trillion times more energy than we have in the highest energy accelerator we have now, the Large Hadron Collider (LHC). To replicate the energies that early in the universe, we'd have to build an accelerator roughly the size of the solar system.

THE FIRST UNIFIED THEORIES

The search for deeper symmetries and unification are very old, going back at least two and a half millennia to Thales of Miletus, although, frankly, probably quite a lot earlier than that. As Francis Bacon described these early unifications:

> Thales asserted Water to be the principle of things . . . that air most plainly is but the exhalation and expansion of water; nay, that even fire itself cannot be lighted, nor kept in and fed, except with moisture and by means of moisture. He saw, too, that the fatness which belongs to moisture, and which is the support and life of flame and fire, seems a kind of ripeness and concoction of the water.

The universe is not made of water. Nor is it made of earth, wind, or fire—the idea put forth by Heraclitus. On the other hand, because all things are made of energy, you could argue that he came the closest if you simply wanted to round fire up to energy, and give him partial credit.

Skipping ahead a few thousand years, by the time Einstein came along, gravity and electromagnetism were the only two forces known, and he wanted nothing more than to unify them into a single theory.

He wasn't the only one.

Just 3 years after Einstein published his final version of General Relativity, a mathematician named Theodor Kaluza developed a new way of thinking about the relationship between gravity and electromagnetism.

Kaluza's approach, expanded and completed by Oskar Klein in 1926, was to write down the equations of general relativity in *five* dimensions—because *why not*? What they found was something quite surprising: The new equations described General Relativity in a universe with three dimensions of space, along with Maxwell's equations of electromagnetism added for good measure.

Kaluza and Klein attempted to unify electromagnetism with gravity by positing that electromagnetism was hidden in an extra dimension.

This isn't as crazy as it seems, but requires a bit of mental gymnastics to make it all come together.

I've made a big deal out of the idea of internal symmetries. Internal symmetries are essentially bookkeeping devices, numbers that you can't actually measure directly, and are swallowed up by a calculation. Phase Symmetry was an internal symmetry, but one with some very specific properties. If you adjust a phase 360 degrees, then you end up where you started. This is *exactly* the defining characteristic of a circle.

The central idea behind the Kaluza-Klein Theory is that what we normally might think of as an internal symmetry is, in fact, an *external* symmetry, a symmetry of the spacetime that we live in. The fourth dimension of space is a Pac-Man dimension in which you loop around and end up where you started.

Given what was known of the physical world at the time, this was a bold approach. There was only one glaringly huge problem: We *don't* live in a universe with four spatial dimensions, and as we saw in Chapter 3, it's a damn good thing that we don't. The only way that all of this can fit together, even approximately, is if the size of the fourth spatial dimension is incredibly tiny, much, much smaller than even the nucleus of an atom.

This is an idea that has resurfaced a lot in the last few decades. Hidden dimensions are the bread and butter of string theory, the current versions of which contain *ten* spatial dimensions, plus one more for time.

The Kaluza-Klein Theory ultimately didn't work out for a number of reasons, the most obvious of which is that the laws of physics *don't* include just gravity and electromagnetism. Also, there's no real place in the theory for quantum mechanics or the existence of photons and gravitons.

The current search for Grand Unified Theories is, in a sense, a step *backward* from those early days. They don't include gravity at all. Instead, the main goal is to unify the strong force with the electroweak.

So far, things aren't really working out for us. But we can still speculate as to what a Grand Unified Theory might look like and, most important, what might it tell us.

WHEN EVERYTHING WAS THE SAME

Each of the fundamental forces responds to a different kind of charge. There is electrical charge, color (for the strong force), weak isospin and hypercharge (for the weak force), and mass (for gravity). But not every particle has every kind of charge. Leptons, for instance, don't have color. Right-handed particles and left-handed antiparticles don't have weak isospin. Photons don't have mass or charge.

If we really want to figure out how everything fits together, we need a good explanation as to why all of these forces and particles look so different today, and how they once could have been so similar.

The earliest Grand Unified Theories were proposed in the early 1970s, shortly after the Standard Model took its current form. One of the first, and most famous, was proposed by Howard Georgi in 1973, and given the rather formidable name SO(10).

The name is just another example of mathematicians' group shorthand. The idea is that all of the particles in the Standard Model are actually just aspects of a single meta-particle. Despite the ten in the name, the SO(10) meta-particle can take one of *sixteen* different forms. From there, working out the details simply involves a lot of counting to show that when you do the math right, there *really* are 16 different particles in the Standard Model.

The up and down quarks each have three possible colors, giving a total of six different quarks in each generation. There is also an electron and a neutrino. Because there's only one color (beige, I guess), we get to add only two to our total, bringing us up to eight particles. Then we simply add the detail that the particles can be either left-handed or right-handed (two possibilities), and we get a total of sixteen different states. Tada!

There are some outstanding features to this theory, and, I must add, to almost any successful Grand Unified Theory (GUT). Because SO(10) envisions the leptons and quarks as just different aspects of the same particle, we get an immediate explanation as to why there are the same number of generations for each, though not why that number is three. The theory also explains the charges of the various particles, including the weird ⅔ for the up quark and –⅓ for the down quark.

But for all of the successes, there are still problems. The weak force is left-handed, and because the neutrino is created only through the weak force, there aren't any right-handed neutrinos. That means that in the actual universe, there are only fifteen particles per generation, not sixteen as originally promised. From this, the theory needs to do a fair amount of mathematical gymnastics, positing that if the right-handed neutrino exists, it's so massive that it won't ever be observed.

SO(10), and lots of other GUTs, predicts that beyond the mediator particles that we've come to know and love, there are also a group of particles called Xs and Ys. These particles allow leptons to turn into quarks and vice versa. This is a very big deal because in the absence of the X particles, it's not clear how we got an excess of protons and neutrons in the first place (remember Chapter 1?).

Moreover, without a way to turn quarks into leptons and vice versa, protons are quite literally immortal. Protons have the distinction of being the lightest baryon, the catchall term for things made of quarks. If you can't get rid of quarks, then as the lightest baryon, protons have nothing to decay into.

But in a GUT, the leptons and quarks are all just different aspects of the same meta-particle, so at very high energies (or with very low probabilities), quarks could spontaneously turn into electrons or vice versa. As a result, the total number of baryons may not be conserved, and protons may not last forever. This turns out to be a *very* good test for your favorite GUT.

For instance, SU(5) is (or was) another very popular GUT.* SU(5) is closely related to SO(10) mathematically and is in many ways much simpler. All things being equal, Occam's razor would suggest that we'll prefer a simpler model over a complicated one. As Georgi and Sheldon Glashow put it in their paper proposing the idea:

> Our hypotheses may be wrong and our speculations idle, but the uniqueness and simplicity of our scheme are reasons enough that it be taken seriously.

On the other hand, SU(5) predicted a proton decay with a lifetime of something like 10^{30} years, which sounds like a hyperbolically large number, until you realize that if we collect lots and lots of protons in the same place, we can actually detect whether any of them decay with that sort of half-life. The current limits are more than 10,000 times longer than that, so SU(5) got thrown out.

This is not the end of the story by any stretch. There are loads of GUTs floating around, some of which predict huge numbers of massive but unobserved particles. Others suggest that the leptons have a *fourth* color beyond the red, blue, and green of the quarks. The biggest experimental guidance is going to come from either actually seeing a proton decay, or perhaps directly detecting a Dark Matter particle.

Meanwhile, nothing prevents us from speculating still further.

AN EXCEPTIONALLY SIMPLE THEORY OF EVERYTHING

Even though we don't yet have an accepted model for a GUT, there are a few scientists who want to cut out the middleman and go straight for

* And, incidentally, discovered by Georgi on the *same night* that he found SO(10).

a Theory of Everything, bringing in gravity as well as the other three forces.

The physicist Garrett Lisi is a story unto himself. He has no academic affiliation, and spends much of his day surfing in Hawaii, and the winter months snowboarding in Colorado. For a good while, he lived out of a car. If you didn't know any better, you might be reluctant to take him seriously, which made it all the more surprising when, in 2007, he put forward a theory that he proposed could explain all of physics. Even more surprising, while the theory hasn't been confirmed, it is extremely elegant.

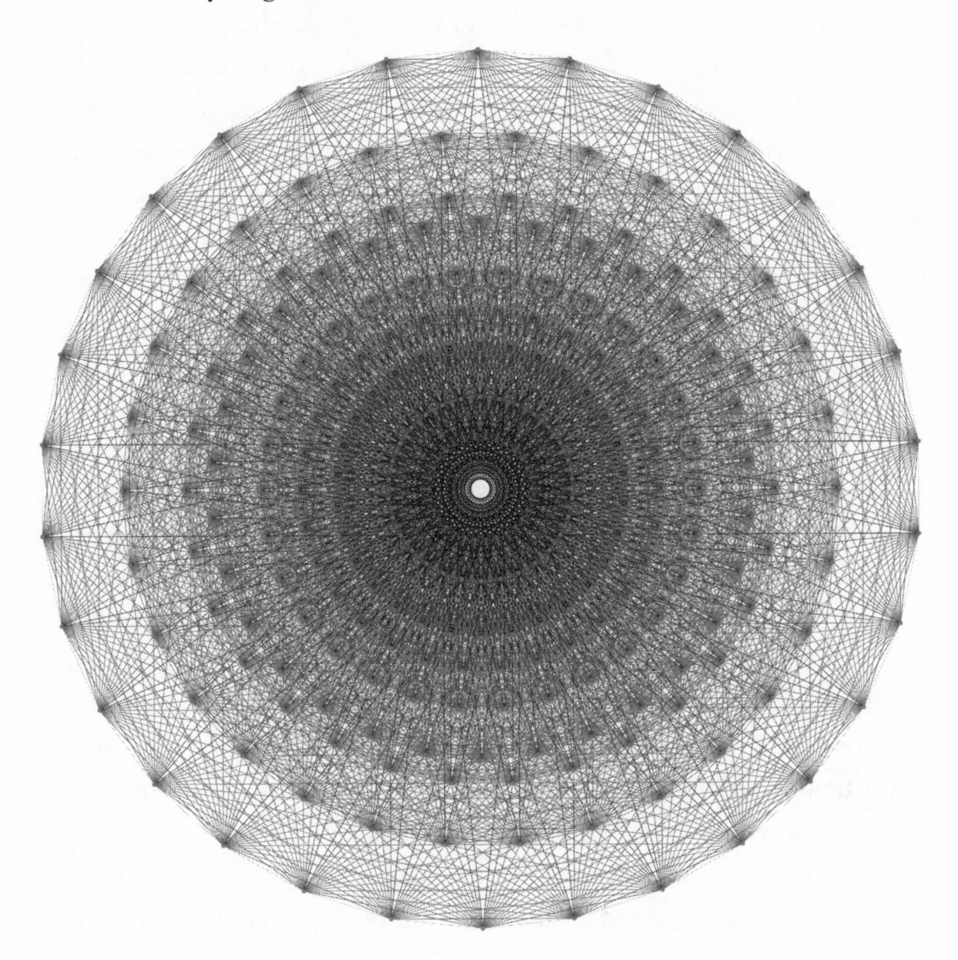

Lisi's theory is based on a mathematical symmetry known as E_8, which he calls "an Exceptionally Simple Theory of Everything." The name itself is actually a mathematician's idea of a joke. *Exceptionally Simple* is not meant to make you think that you're a dummy for not understanding it. Rather, E_8 is one of five special Exceptionally Simple Symmetries identified by the awesomely named mathematician Wilhelm Killing in the late nineteenth century.

The image on the previous page is just one way of looking at the E_8 Symmetry. Every point represents a particle, but just as with our diagram illustrating the Standard Model, two quarks with different colors and different spins count as different points on the diagram.

I don't expect you to count up all possible particles in the Standard Model and on the E_8 diagram, but if you were to do so, you'd come to the realization that there are twenty-two particles missing and that filling in those particle holes makes predictions for things that we haven't yet seen.

One of the things that makes E_8 interesting is that it includes gravity as well as the strong force. As a result, mass, which plays the equivalent role of charge for gravity, is included as a separate point for each of the three generations.

Lisi himself has described his theory as a work in progress, and I should say, lest there be any confusion on this count, that even though a Theory of Everything contains gravity, it is *not* the same thing as a theory of quantum gravity. In other words, under the best of circumstances, E_8 still won't be able to deal with strong gravitational fields.

There has been enormous skepticism within the physics community. For one thing, the model doesn't actually predict the masses of second- and third-generation particles, even if you put in the masses of the first-generation particles (the particles most familiar to us) by hand.

For another, the E_8 theory is very cavalier about combining fermions and bosons into a single meta-particle. This is a big deal. Remember that you have to turn fermions around twice to have them look the same as when you started, whereas most bosons will look the same after a single rotation. The two don't trivially combine, because doing so would seem to be a violation of symmetry.

That said, the relationship between fermions and bosons may yet lead to one of the most important steps forward in our understanding of symmetry in the universe.

SUPERSYMMETRY

We have a pile of particles called fermions and another called bosons, and they interact with one another when a force is needed, after which they call it a day. Organizationally, they seem quite different. The fermions are arranged into neat little rows and columns, and the bosons are strewn everywhere, a consequence of whatever haphazard symmetry gave rise to them.

But they also play off each other in surprising ways.

Consider the workings of the Higgs boson. Just as the Higgs gives rise to the masses of particles, other particles should also contribute to the mass of the Higgs boson itself. As with the Higgs mechanism, it's all a matter of how one field interacts with another. Because *interaction* is simply a fancy word for "energy," and energy and mass are interchangeable; the Higgs mass that we measure at the LHC isn't necessarily the real mass that it would have if we could strip away all of those interactions. This is exactly analogous to that whole issue of shielding and the bare charge of the electron. What you see is not necessarily what you get.

And the correction to the Higgs mass should be *huge*, generally of the order of the Planck mass. The folks at the LHC measured a Higgs mass about 200 quadrillion times smaller than the Planck mass.

The fact that the mass is so small, but not exactly zero, is too much to accept by chance. This means that the true mass of the Higgs would have to be incredibly finely tuned so that the correction and the bare mass *almost* (but don't exactly) cancel each other to about 1 part in 10^{17}.

The odds of something like that happening in nature by mere chance is so remote as to be laughable.

I gave you the correction only for electrons and positrons, but there are lots of other types of particles out there. Each and every one is going to interact with the Higgs and add a correction to the mass.

There's a weird wrinkle to all of this. We saw earlier that fermions are associated with a −1, and bosons got a +1 when you switched two identical particles.

Those plus and minus 1s are going to be drafted into service again; they just play a slightly different role this time around. For each species of fermion, we *subtract* from the bare mass to get the observed mass—that's why I subtracted when talking about electrons—and with bosons we *add* to the bare mass. Wouldn't it be nice (I ask only semirhetorically) if there were the same number of types of fermions as bosons? Under the best of circumstances, they might exactly cancel.

We don't actually *care* what the bare mass is, like we don't care what the bare electric charge is. It's just deeply unnerving that the corrections are so *precise.* But if we want to bury our heads in the sand, we're allowed to.

Richard Feynman did not like this (even though he was forced to use it):

> The shell game that we play . . . is technically called "renormalization." But no matter how clever the word, it is still what I would call a dippy process! Having to resort to such hocus-pocus has prevented us from proving that the theory of quantum electrodynamics is mathematically self-consistent. It's surprising that the theory still hasn't been proved self-consistent one way or the other by now; I suspect that renormalization is not mathematically legitimate.

When chemists do experiments on molecules, they can almost always ignore the atomic nature of matter. When atomic physicists work on atoms, they almost always ignore the interactions of quarks. You could just as easily think of a psychologist ignoring the detailed neurochemical interactions in the brain. While it's certainly true that a complete enough knowledge of the brain would tell us something useful about behavior, it's not as though a psychologist can't make useful conclusions without all of that low-level detail.

While we can't measure the bare mass of the Higgs, it *is* strange that it ends up being small enough that we can actually detect it. How does everything cancel so nearly perfectly? This sort of fine-tuning correction shows up all the time in physics. There needs to be some better reason for everything working out so perfectly beyond "that's how it is."

One of the best attempts at an explanation comes from an idea known as supersymmetry (as SUSY to its friends). Grand Unified Theories unify all of the fermions (the particles of matter) into a single particle, and the bosons into, essentially, a single force, but SUSY takes this a step further. SUSY supposes that even bosons and fermions are just different sides of the same coin. For every boson, there should be a fermion and vice versa. This is more complicated than it first appears, as illustrated by the difficulties with the E_8 theory. Fermions and bosons really are quite different beasts.

Besides, in the Standard Model at least, there *aren't* equal numbers of fermions and bosons. Including all of the combinations of spin and color, there are twenty-eight different bosons and ninety different fermions.

No matter. The solution is simply to hypothesize more particles. *Every* particle gets a partner of the opposite type. An electron is a fermion. On the other side is a boson called a *s*electron. A photon is a boson. It gets a fermion partner called the phot*ino*, and so forth.*

* Just so you don't embarrass yourself when talking about this at your next party, the partner of the W, the wino, is pronounced *weeno.*

I realize that the solution "make up a bunch of new particles" sounds (1) so easy that you don't need an advanced physics degree to come up with it, and (2) so silly that it's not clear that it'll do anything. But bear with me for a moment. First off, the idea of coming up with symmetries, in this case between fermions and bosons, is really important in physics. The way we understand the weak and electromagnetic forces is ultimately by supposing that the electron and the neutrino (also the up and down quarks) are just different aspects of the same fundamental particle. It's this symmetry that ultimately gives rise to our understanding of the Higgs.

If every particle gets a partner, it does seem strange that we've never seen one, doesn't it? Maybe.

One of the generic predictions of supersymmetric models is that the supersymmetric partners should be hundreds or even thousands of times larger than our familiar versions of them. And very massive particles, as you know, don't stick around for long.

There may be a whole bunch of particle states called neutralinos, which are (as you might guess) electrically neutral. This means that even if we were to make them in an accelerator, they would be very, very tough to detect directly. Essentially, we'd have to look for electron–positron or muon–antimuon pairs with a ton of missing energy. That missing energy would be the neutralino sneaking out of the particle detector unnoticed like a thief in the night.

I should warn you now that the initial experimental results from the LHC as well as experiments to detect SUSY particles directly don't look terribly promising. Although there are lots of possible SUSY models, many of them fall into a group called the Minimal Supersymmetric Standard Model (MSSM), most versions of which suggest that we should have detected supersymmetry already, if it exists. There's only a small range of masses in which SUSY particles could be hiding that we haven't yet checked, although particles are always in the last place you look for them. We had a similar sort of situation with the Higgs.

It'd be a shame if SUSY turned out to be wrong, because it would give us huge hints about a lot of outstanding problems. For instance, the lightest supersymmetric partner (the lightest neutralino, probably) would still be pretty massive, but it would be able to fly in and out of particle accelerators unnoticed.

Hmm . . . a massive, abundant particle that's stable because there's nothing for it to decay into? Sounds like Dark Matter. If only SUSY turns out to be a real property.

In case you hadn't already guessed, I'm ardently hoping against hope that sypersymmetry turns out to work, but the part of my brain that is informed by experimental evidence warns that this is looking like a bad bet.

Even if supersymmetry is a fact of our universe, it must be at least a little bit broken. If it weren't, all the partners would be the same mass as the originals. But if *that* were the case, we would have discovered them long ago.

One final note. Supersymmetry is often tied to string theory, and in particular, people will talk about *superstrings*, for the simple reason that string theory and its load of extra dimensions requires supersymmetry as part of the model. The converse does not hold. Supersymmetry could well be right without string theory.

BEYOND SYMMETRY

Imagine row after row of identical spinning tops.

Eventually, of course, one of the tops will stumble, *Inception*-style, into its neighbor. Which top and which direction are entirely random, but once the top stumbles, the symmetry is broken forever. Moreover, as the tops collapse, the patterns are far more complicated than you might have predicted given the initial simplicity of the system.

Provided we had the patience to keep setting it up, we could run this scenario again and again, and each time the tops would fall in a different way. If we simply imagine the pattern of collapse as representing

Tops and Symmetry Breaking

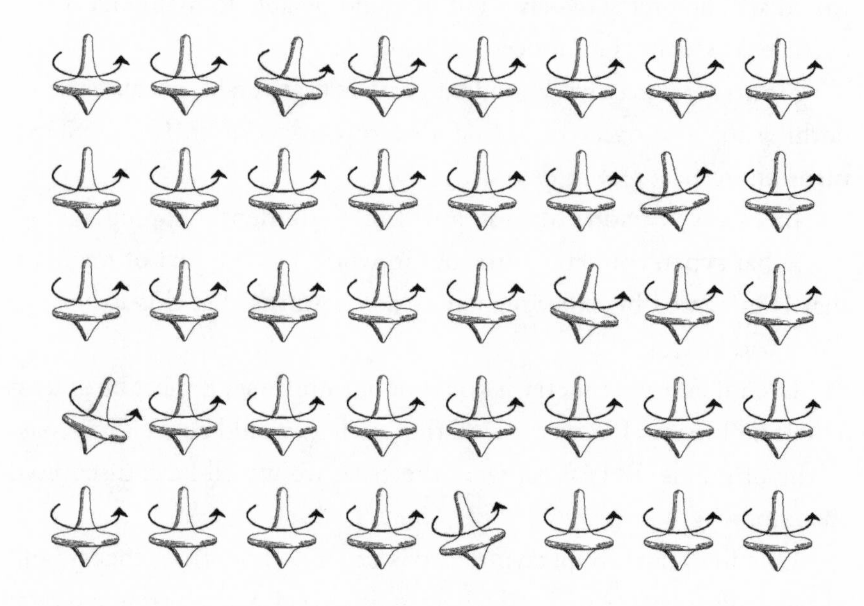

the evolution of physics in the early universe, it's almost as if we're able to explore various regions of the multiverse.

The tops start out symmetrically, but they end up in apparent disarray. The universe is the same way.

We started our discussion of symmetry in the world of the abstract, by talking about circles and polyhedrons and the like. This wasn't just so that we could draw lots of cool pictures; the laws of nature seem to be written symmetrically. But even if nature starts off symmetrically, it doesn't end up that way. Somehow, symmetry isn't, can't be, the end of the story.

Our entire picture of how the universe works is based on an interplay between symmetry and randomness, and to be perfectly frank, we're not entirely sure where one ends and the other begins. Randomness—or chaos, really—is often the signifying characteristic of a bad guy, while the orderly (symmetric) character is the hero. This is entirely unfair.

Although I don't want to get *too* touchy-feely, there's an important element of truth to philosophies that embrace the yin and the yang.

The laws of the universe are symmetric, but once we introduce the demon of randomness, the results of those laws, the universe we see around us, are most definitely *not* going to appear symmetric. Randomness is the hallmark of a quantum mechanical universe. Start with the same initial configuration and run an experiment again and again, and you'll get different, perhaps profoundly different, results.

We've seen this time and time again, from the bumps and wiggles of structure in the large-scale universe down to the symmetry breaking of the Higgs.

Randomness can conceal but not entirely erase the hidden symmetry of the universe. In the end we don't know how much randomness dictates the nature of the apparently broken symmetries that define the laws we see today, but we do know that even once the (still fairly symmetric) laws were in effect, randomness continued to blur our view of what would otherwise be a very orderly universe. It was randomness that first gave rise to structure, even though the symmetric laws of gravity allowed that structure to grow, collapse, form into stars and build complex chemicals (and, I might add, life). It's randomness that governs radioactivity, the fusion in the sun, and very likely the neurons in our brains.

In the end, broken symmetry—the very fact that the universe in the rearview mirror is different from our own—is ultimately what gives rise to a universe interesting enough to live in.

ROADSIDE ATTRACTION 1: THE PARTICLE ZOO

THE FERMIONS (SPIN-½)

	Quarks		Leptons	
First generation	Up	Down	Electron neutrino	Electron
Second generation	Charmed	Strange	Mu neutrino	Muon
Third generation	Top	Bottom	Tau neutrino	Tau

Increasing Mass ↓

THE BOSONS (SPIN-0, SPIN-1, SPIN-2?)

Particle	Spin	What it does	Does it exist?
Higgs	0	Gives mass to W, Z, quarks, and charged leptons	Seems to (2012), Large Hadron Collider
Photon	1	Mediator of the electromagnetic force (a.k.a. light)	Yup, photoelectric effect (1905)
W^+, W^-, Z^0	1	Mediators of the weak force	Yes, discovered at the Super Proton Synchrotron (1983)

Gluons	1	Mediators (there are eight different color combinations) of the strong force.	Yes, discovered by the PLUTO experiment (1978)
Gravitons	2	Mediator of gravity	Who knows? We don't even know how they would work.

A FEW IMPORTANT COMPOSITE PARTICLES

Proton: up + up + down

Neutron: up + down + down

Neutral pion up + anti-up *or* down + anti-down

Neutral kaon down + anti-strange *or* strange + anti-down

ROADSIDE ATTRACTION 2: A SIDE OF SYMMETRIES

The following list is *not* comprehensive, but it should serve as a useful reference throughout this book and (should you care to work it into conversation) in life.

Noether's First Theorem: Every continuous symmetry corresponds to a conserved quantity.

DISCRETE SYMMETRIES IN PHYSICS

C Symmetry (for charge conjugation): The physical laws for antiparticles behave *exactly the same* as for their respective particles. It's a symmetry for all of the forces but the weak force.

P Symmetry (for parity): All of the laws of physics look just as valid if you view everything through a mirror. Also broken for the weak force.

T Symmetry (for time reversal): The laws of physics look the same upon a reversal of the flow of time. By itself, this one is *also* broken for the weak force.

CP Symmetry: Combination of C and P. This is *also* violated in the weak force, and a good thing too, because without a CP violation, there'd be no excess matter in the universe.

CPT Symmetry: Combination of C, P, and T, applied in any order. A perfect symmetry of our universe, so far as we can tell.

Particle Replacement Symmetry: All of the measurable quantities in a system will remain unchanged if you swap two particles of identical type and state. The complication is that if you exchange two fermions, the wave function is multiplied by a –1, but this isn't anything you can detect directly.

CONTINUOUS SYMMETRIES IN PHYSICS

Time Translation Symmetry: All of the laws of physics behave the same at different times. From Noether's Theorem, this leads to conservation of energy.

Translational Symmetry: The laws of physics are exactly the same at all places in the universe. On large scales, this is reflected in the homogeneity of the universe and the Cosmological Principle. From Noether's Theorem, this leads to conservation of momentum.

Rotational Symmetry: The laws of physics don't change if you rotate a system in its entirety. On large scales, this is reflected in the isotropy of the universe and is also part of the assumptions underlying the Cosmological Principle. This leads to conservation of angular momentum.

Lorentz Invariance: The laws of physics look equally valid for any observer moving at constant speed and direction. It's also the basis for Special Relativity.

Weak Equivalence Principle: Particles moving in free fall are locally indistinguishable from inertial systems. This serves as the basis for General Relativity.

GAUGE SYMMETRIES (WITH SCARY MATH NOTATION)

Phase Symmetry, U(1): The phase of a field can change and there will be no measurable effect.

Electron–Neutrino Symmetry, $SU(2)_L$: The weak interaction behaves exactly the same if you switch all of your neutrinos to electrons and vice versa, so long as they're all left-handed.

Color Symmetry, SU(3): The strong interaction behaves exactly the same on red, green, or blue quarks. Switch one color for another (so long as you do it consistently), and the force is exactly the same.

SU(5): One of the earliest and most well tested Grand Unified Theories. It predicts extra particles that lead to proton decays. It's been ruled out by the observed proton lifetime.

SO(10): A current, popular Grand Unified Theory which predicts a long proton lifetime as well as an extremely massive right-handed neutrino.

E_8: A speculative Theory of Everything that purports to explain the masses of the various generations of particles, why there are three of them, and also gravity.

SUSY: The nickname for supersymmetry, which says that every fermion has a boson partner, and every boson has a fermion. Early experimental results are not looking good.

FURTHER READING

POPULAR WORKS ON PHYSICS AND SYMMETRY (WHICH ALL SCIENCE GEEKS SHOULD CONSUME)

Abbott, Edwin. *Flatland: A Romance of Many Dimensions*. New York: Dover, 1992. The adventures of "A Square" in a strongly hierarchical, geometrical universe.

Bryson, Bill. *A Short History of Nearly Everything*. New York: Broadway Books, 2003. A wonderful tour of the history of science, including some outstanding anecdotes.

"Desperately Seeking Symmetry." *Radiolab*. WNYC Radio, April 18, 2011.

Du Sautoy, Marcus. *Symmetry: A Journey into the Patterns of Nature*. New York: Harper, 2008. This is a bit more autobiographical, but provides a good basis for understanding symmetries in nature and in mathematics.

Gardner, Martin. *The New Ambidextrous Universe: Symmetry and Asymmetry from Mirror Reflections to Superstrings*. 3rd rev. ed. Mineola, NY: Dover, 2005. A classic (semitechnical) work on symmetry in mathematics, nature, and physical law.

Goldberg, Dave, and Jeff Blomquist. *A User's Guide to the Universe: Surviving the Perils of Black Holes, Time Paradoxes, and Quantum Uncertainty*. Hoboken, NJ: Wiley, 2009. A fun romp through space and time by your humble author.

Lederman, Leon M., and Christopher T. Hill. *Symmetry and the Beautiful Universe*. Amherst, NY: Prometheus Books, 2004. An excellent discussion of how symmetry plays out in fundamental physics.

Stewart, Ian. *Why Beauty Is Truth: The History of Symmetry.* New York: Basic Books, 2007.

Weyl, Hermann. *Symmetry.* Princeton, NJ: Princeton University Press, 1952. Weyl did more than almost anyone to help promote the role of symmetry in physics. This work is a classic and includes discussions not only of physics but also of tilework, philosophy, and history.

BIBLIOGRAPHY

INTRODUCTION

Anderson, P. W. "More Is Different." *Science* 177, no. 4047 (1972): 393–396. Quote from the introduction.

Feynman, Richard Phillips, Robert B. Leighton, and Matthew L. Sands. "Basic Physics." In *The Feynman Lectures on Physics*. Redwood City, CA: Addison-Wesley, 1989, p. 2-2. Feynman described his chess analogy a number of times, but there are few better ways of thinking of exactly what physics is trying to get at.

Feynman, Richard P., Robert B. Leighton, and Matthew L. Sands. "Symmetry in Physical Laws." In *The Feynman Lectures on Physics*. Reading, MA: Addison-Wesley, 1963, 1965, pp. 52-1-12. This is one of the best introductions to general physical symmetries (as known at the time) that I've seen. If you can get your hands on it, I would strongly recommend listening to the audiobook version.

Galilei, Galileo. *Two New Sciences*. Dover, 1914. Among much else, Galileo discusses the prospects of giants and shows that giant bones would need to be unreasonably thick to provide the necessary support.

Haldane, J. B. S. "On Being the Right Size." *Harper's Magazine*, March 1926. Haldane gives the rundown of why insects are more or less immune to gravity but not to surface tension.

Horgan, John. *The End of Science: Facing the Limits of Knowledge in the Twilight of the Scientific Age*. Reading, MA: Addison-Wesley, 1996.

Scoular, Spencer. *First Philosophy: The Theory of Everything*. Boca Raton, FL: Universal, 2007. This work contains a number of excellent quotes on the nature of symmetry.

Stannard, Russell. *The End of Discovery*. Oxford: Oxford University Press, 2010.

CHAPTER 1

Birks, J. B. *Rutherford at Manchester*. New York: Benjamin, 1962. Origin of the "All science is either physics or stamp collecting" quote.

Brown, Dan. *Angels & Demons*. New York: Atria Books, 2000.

Carroll, Lewis. *Through the Looking-Glass, and What Alice Found There*. London: Macmillan, 1871. A classic of silliness, symmetry, and mathematical humor. Also, a big inspiration for the present volume.

ChemSpider: The Free Chemical Database. www.chemspider.com. A website run out of the Royal Society of Chemistry listing all known compounds, approximately 26 million, as of this writing.

Chown, Marcus. *The Magic Furnace: The Search for the Origins of Atoms*. Oxford: Oxford University Press, 2001. This provides a gripping description of, among other things, Marie Curie's lead-lined box of papers and her radioactive fingerprints.

Close, F. E. *Neutrino*. Oxford: Oxford University Press, 2010. A very good description of the history of the neutrino and the discovery of neutrino oscillations and neutrino mass.

Fukuda, Y., T. Hayakawa, E. Ichihara, et al. "Measurements of the Solar Neutrino Flux from Super-Kamiokande's First 300 Days." *Physical Review Letters* 81, no. 6 (1998): 1158–1162. A technical paper describing the first experimental detection of the existence of neutrino mass.

Hofstadter, Douglas R. *Gödel, Escher, Bach: An Eternal Golden Braid*. New York: Basic Books, 1979. Hofstadter's classic touches on an enormous number of symmetries. In particular, he describes J. S. Bach's "crab canon," a musical palindrome.

James, Laylin K. *Nobel Laureates in Chemistry, 1901–1992*. Washington, DC: American Chemical Society, 1993.

Rutherford, E. "The Scattering of α and β Particles by Matter and the Structure of the Atom." *Philosophical Magazine* 21 (1911): 669–688. One of the great landmark papers in physics. Rutherford does nothing less than discover the nature of atoms.

Tuniz, Claudio. *Radioactivity: A Very Short Introduction.* Oxford: Oxford University Press, 2012.

Wald, G. "The Origin of Optical Activity." *Annals of the New York Academy of Science* 69, no. 2 (1957): 352–368. Quote from Einstein regarding matter and antimatter and how matter "won the fight."

Xue, L., and the STAR Collaboration. "Observation of the Antimatter Helium-4 Nucleus at the RHIC." *Journal of Physics G: Nuclear and Particle Physics* 38, no. 12 (2011): 124072.

CHAPTER 2

Albert, David Z. *Time and Chance.* Cambridge: Harvard University Press, 2000. An interesting discussion of the philosophy of thermodynamics and the origin of the Past Hypothesis.

Callender, Craig. "Thermodynamic Asymmetry in Time." In *The Stanford Encyclopedia of Philosophy.* Ed. Edward N. Zalta. Fall 2011. http://plato.stanford.edu/archives/fall2011/entries/time-thermo.

Callender, Craig. "There Is No Puzzle about the Low Entropy Past." In *Contemporary Debates in the Philosophy of Science.* Ed. C. Hitchcock. Oxford: Blackwell, 2004, chap. 12. Callender describes a scenario in which a prophesy is made (and fulfilled) requiring all sorts of improbable events. The universe, as run in reverse, seems to be going through similarly unlikely chains of events to reach the very low entropy state of the early universe.

Carroll, Sean M. *From Eternity to Here: The Quest for the Ultimate Theory of Time.* New York: Dutton, 2010. An interesting speculative voyage into the possibility that entropy gives rise to the arrow of time rather than the other way around.

Clausius, Rudolf. Second Law of Thermodynamics described in Poggendorff's *Annalen der Physik* 93 (December 1854): 481. Available in English: *Journal de Mathematiques* 20 (1855) and *Philosophical Magazine* 12, s. 4 (August 1856): 81. One of the earliest descriptions of the Second Law of Thermodynamics, and functionally, the one we use today.

Feynman, Richard P. "Nobel Lecture." Nobelprize.org. December 11, 1965. www.nobelprize.org/nobel_prizes/physics/laureates/1965/feynman-lecture.html.

Hoyle, F. "A New Model for the Expanding Universe." *Monthly Notices of the Royal Astronomical Society* 108 (1948): 372. Hoyle was one of the main proponents of a steady, state model in which the universe continuously created matter to "fill in the gaps."

Isaacson, Walter. *Einstein: His Life and Universe.* New York: Simon & Schuster, 2007, pp. 254–255. Isaacson does a masterful job in describing both the state of the science when Einstein developed relativity, as well as (in chap. 15) his recanting of the cosmological constant upon the discovery of the expanding universe.

Landauer, Rolf. "Irreversibility and Heat Generation in the Computing Process." *IBM Journal of Research and Development* 5 (1961): 183–191.

Lovelock, James. *The Ages of Gaia: A Biography of Our Living Earth.* New York: Norton, 1988. An excellent account of the physics and physiology of the Oklo site. Oddly, for such a significant natural experiment, there is remarkably little written about it.

Steinhardt, Paul J., and Neil Turok. "Cosmic Evolution in a Cyclic Universe." *Physical Review D* 65, no. 12 (May 24, 2002): 126003. The Ekpyrotic Universe is one of several cyclic models of the universe.

Verlinde, E. "On the Origin of Gravity and the Laws of Newton." *Journal of High Energy Physics* 4, no. 29 (2011): 1-27. Verlinde claims that gravity is emergent from the Second Law of Thermodynamics.

White, T. H. *The Once and Future King.* New York: Putnam, 1958. White's version of Merlyn portrayed him as an old man who lived backward.

CHAPTER 3

Adams, Douglas. *The Hitchhiker's Guide to the Galaxy.* New York: Ballantine Books, 1980.

Aristotle. "Metaphysics." In *A New Aristotle Reader.* Ed. J. L. Ackrill. Princeton, NJ: Princeton University Press, 1987, pp. 255–360.

Aristotle. "Physica." In *The Works of Aristotle.* Eds. W. D. Ross, Robert P. Hardie, and R. K. Gaye. Oxford, UK: At the Clarendon, 1930. A genius of the first order, Aristotle was nevertheless wrong about just about everything he said about the physical world.

Barrow, John D. "Cosmology, Life, and the Anthropic Principle." *Annals of the New York Academy of Sciences* 950, no. 1 (2001): 139–153.

Bruno, Giordano. "5th Dialogue." In *Cause, Principle and Unity.* Eds. Richard J. Blackwell and Robert de Lucca. Cambridge: Cambridge University Press, 1998.

Bryson, Bill. *A Short History of Nearly Everything*. New York: Broadway Books, 2003. Bryson has a very nice description of the bet between Halley, Hooke, and Wren that led ultimately to Newton's publication of the *Principia*.

Copernicus, Nicolaus. *On the Revolutions of the Heavenly Spheres*. Ed. A. M. Duncan. Newton Abbot, UK: David & Charles, 1976.

Dyson, F. J. "Search for Artificial Stellar Sources of Infrared Radiation." *Science* 131, no. 3414 (1960): 1667–1668. Dyson is one of the greatest living physicist-futurists, and his suggestion of what has become known as a Dyson sphere (first presented in this paper) remains one of the staples of both science fiction and the search for extraterrestrial intelligence.

The Extrasolar Planets Encyclopaedia Catalog. http://exoplanet.eu/catalog.php. As of this date, the encyclopedia lists 716 candidate planets.

Feynman, Richard P. *QED: The Strange Theory of Light and Matter*. Princeton, NJ: Princeton University Press, 1985, p. 10. This is the origin of his quote regarding the fine structure constant.

Galilei, Galileo. *Discoveries and Opinions of Galileo: Including The Starry Messenger (1610), Letter to the Grand Duchess Christina (1615), and Excerpts from Letters on Sunspots (1613), The Assayer (1623)*. Ed. Stillman Drake. New York: Anchor, 1990.

Gamow, George. *The Great Physicists from Galileo to Einstein*. New York: Dover, 1961. This is an excellent history of some of the formative figures in physics. In particular, Gamow points out that it was Aristotle who coined the word *physics* (p. 5).

Gould, Stephen Jay. "The Late Birth of a Flat Earth." In *Dinosaur in a Haystack: Reflections in Natural History*. New York: Harmony, 1995.

Harrison, E. *Darkness at Night: A Riddle of the Universe*. Cambridge: Harvard University Press, 1987.

Heath, Thomas. *Aristarchus of Samos: The Ancient Copernicus*. New York: Dover, 1981.

Kepler, Johannes. *New Astronomy*. Ed. William Halsted Donahue. Cambridge: Cambridge University Press, 1992.

"Kepler: A Search for Habitable Planets." NASA. www.nasa.gov/mission_pages/kepler/main/index.html.

Kepler, Johannes. *Mysterium Cosmographicum = The Secret of the Universe*. Ed. E. J. Aiton. New York: Abaris, 1981.

Koestler, Arthur. *The Sleepwalkers: A History of Man's Changing Vision of the Universe.* London: Hutchinson, 1959. Koestler describes a number of interesting incidents involving Tycho and Kepler, including one I left out involving Tycho's drunken pet moose falling down a staircase and dying. He also provides background for the various quotes by Kepler.

Land, Kate, and João Magueijo. "Examination of Evidence for a Preferred Axis in the Cosmic Radiation Anisotropy." *Physical Review Letters* 95, no. 7 (2005): 71301. The origin of the "Axis of Evil" in the cosmic microwave background.

Ptolemy. *Almagest.* Trans. G. J. Toomer. New York: Springer-Verlag, 1984.

Shapley, H. "Globular Clusters and the Structure of the Galactic System." *Publications of the Astronomical Society of the Pacific* 30 (1918): 42.

Tegmark, Max. "The Multiverse Hierarchy." In *Universe or Multiverse?* Ed. Bernard Carr. Cambridge: Cambridge University Press, 2007. This entire volume is packed with interesting discussions on the nature of the multiverse. Tegmark contributed a hierarchy to describe how *multiverse* can mean many different things.

Tegmark, Max. "On the dimensionality of spacetime." *Classical and Quantum Gravity* 14, no. 4 (1997): L69–L75.

Webb, J. K., J. A. King, M. T. Murphy, et al. "Indications of a Spatial Variation of the Fine Structure Constant." *Physical Review Letters* 107, no. 19 (2011): id. 191101.

CHAPTER 4

Angier, Natalie. "The Mighty Mathematician You've Never Heard of." *The New York Times*, March 27, 2012. A nice biography of Noether on the occasion of her 130th birthday.

Brewer, James W., and Martha K. Smith. *Emmy Noether: A Tribute to Her Life and Work.* New York: Dekker, 1981.

Dick, August. *Emmy Noether, 1882–1935.* Trans. H. I. Blocher. Birkhauser: Boston, 1981. This work gives, among much else, Weyl's complete memorial address.

Goldberg, David M., and J. Richard Gott. "Flexion and Skewness in Map Projections of the Earth." *Cartographica: The International Journal for Geographic Information and Geovisualization* 42, no. 4 (2007): 297–318. Apropos of the discussion of the relation between map projections and great circle routes, Rich Gott and I came up with a systematic approach to describing the distortions on large scales. That is, how far will you have to deviate from a straight line on the map to follow a great circle?

Hamilton, W. R. "On a General Method in Dynamics." *Philosophical Transactions of the Royal Society* Part II (1834): 247–308; Part I (1835): 95–144. In *Sir William Rowan Hamilton (1805–1865): Mathematical Papers*. Ed. David R. Wilkins. Dublin: School of Mathematics, Trinity College, 2000.

Huygens, Christiaan. *Treatise on Light, in Which Are Explained the Causes of That Which Occurs in Reflexion, and in Refraction and Particularly in the Strange Refraction of Iceland Crystal*. Reprint ed. New York: Dover, 1962.

Maupertuis, Pierre-Louise Moreau. "Derivation of the Laws of Motion and Equilibrium from a Metaphysical Principle." 1746. Wikisource. en.wikisource.org/w/index.php?title=Derivation_of_the_laws_of_motion_and_equilibrium_from_a_metaphysical_principle&oldid=2169853.

Neuenschwander, Dwight E. *Emmy Noether's Wonderful Theorem*. Baltimore, MD: Johns Hopkins University Press, 2010. An excellent semitechnical discussion of Noether's Theorem and much of the variational mechanics that led up to it.

Newton, Isaac. *The Principia*. Trans. I. B. Cohen and A. Whitman. Berkeley: University of California Press, 1999. This was the source for my translations of Newton's laws of motion.

Noether, Emmy. "Invariante Variationsprobleme." *Nachrichten von der Gesellshaft der Wissenshaften zu Göttingen, Math-phys. Klasse* (1918): 235–257. Trans. M. A. Tavel. http://arxiv.org/abs/physics/0503066v1.

Pickover, Clifford A. *Archimedes to Hawking: Laws of Science and the Great Minds Behind Them*. Oxford: Oxford University Press, 2008. Pickover provides the outstandingly boastful quote from Bernoulli in which he challenges other mathematicians to solve the Brachistochrone problem.

Singh, Simon. *Fermat's Enigma: The Epic Quest to Solve the World's Greatest Mathematical Problem*. New York: Walker, 1997. This is a nice description of both Fermat's Last Theorem and Andrew Wiles's solution.

Smolin, Lee. "On 'Special Relativity: Why Can't You Go Faster Than Light?' by W. Daniel Hillis." *Edge* 52 (March 28, 1999). http://edge.org/documents/archive/edge52.html.

CHAPTER 5

Adam, T., N. Agafonova, A. Aleksandrov, et al. "Measurement of the Neutrino Velocity with the OPERA Detector in the CNGS Beam." http://arxiv.org/

abs/1109.4897v2. This preprint sparked enormous interest in the possibility of a faster-than-light neutrino. Ultimately, the result was shown to be flawed and the result of loose cables.

Card, Orson Scott. *Ender's Game*. New York: Tor, 1991. One of the central plot points involves the use of an ansible to direct an interstellar armada.

Cartlidge, Edwin. "Leaders of Faster-Than-Light Experiment Step Down." *Science*, March 30, 2012. http://news.sciencemag.org/scienceinsider/2012/03/leaders-of-faster-than-light-exp.html. Shortly after an apparently mechanical explanation for the supposed faster-than-light neutrinos, the OPERA spokesperson, Antonio Ereditato, and the physics coordinator, Dario Autiero, stepped down from their leadership positions in the experiment.

Einstein, Albert. "Does the Inertia of a Body Depend on Its Energy Content?" In *The Principle of Relativity*. Trans. George Barker Jeffery and Wilfrid Perrett. London: Methuen, 1923.

Einstein, Albert. "On the Electrodynamics of Moving Bodies." In *The Principle of Relativity*. Trans. George Barker Jeffery and Wilfrid Perrett. London: Methuen, 1923.

Einstein, Albert. "Über die vom Relativitätsprinzip geforderte Trägheit der Energie." *Annalen der Physik* 328, no. 7 (1907): 371–384. Einstein recognized that sending a signal faster than light could mess up causality.

Frisch, D. H., and J. H. Smith. "Measurement of the Relativistic Time Dilation Using μ-Mesons." *American Journal of Physics* 31, no. 5 (1963): 342–355.

Galilei, Galileo. *Dialogue Concerning the Two Chief World Systems, Ptolemaic & Copernican*. Trans. Stillman Drake. Berkeley: University of California, 1953.

Isaacson, Walter. *Einstein: His Life and Universe*. New York: Simon & Schuster, 2007, pp. 107–139. Isaacson describes the various contradictory statements that Einstein made regarding his knowledge (or lack thereof) of the Michelson–Morley experiment and its subsequent influence (or not) on his development of Special Relativity.

Le Guin, Ursula K. *Three Hainish Novels*. Garden City, NY: Nelson Doubleday, 1966. This work, and in particular *Rocannon's World*, introduced the word *ansible* as a theoretical device that can communicate faster than light.

Minkowsi, H. "Space and Time." In *The Principle of Relativity: A Collection of Original Memoirs on the Special and General Theory of Relativity*. A. Einstein, H. A. Lorentz, H. Weyl, and H. Minkowski. New York: Dover, 1952, pp. 109–164. A reprint of the original 1923 edition.

Roddenberry, Gene. Letter. *Sky and Telescope,* June 1991. In this letter, Roddenberry, the creator of *Star Trek,* endorses the idea that the Vulcan home world is in orbit around 40 Eridani A.

Rossi, B., and D. B. Hall. "Variation of the Rate of Decay of Mesotrons with Momentum." *Physical Review* 59, no. 3 (1941): 223–228. A description of the first experiment to measure that the number of muons measured at ground level were only a little smaller than the counts at the top of a mountain. If time dilation weren't real, the muons (called mesotrons in the title of the article, because they weren't understood at the time) should basically have all decayed before they hit the ground.

Tolman, R. C. "Velocities Greater Than That of Light." In *The Theory of the Relativity of Motion*. Berkeley, CA: University of California Press, 1917, p. 54. This work, written not too long after Einstein developed relativity, is the origin of the tachyonic antitelephone.

Viereck, George. "What Life Means to Einstein: An Interview by George Sylvester Viereck." *The Saturday Evening Post,* October 26, 1929, p. 17. This is the origin of the quote relating the difficulties of thinking about spacetime in four dimensions.

CHAPTER 6

Casimir, H. B. G. "On the Attraction between Two Perfectly Conducting Plates." *Proceedings of the Koninklijke Akademie van Wetenschappen te Nederland* (1948): 793. This is one of the first descriptions of what has come to be known as the Casimir effect, one of the strongest pieces of experimental evidence that there is a real vacuum energy.

Einstein, Albert. "Die Relativitäts-Theorie." *Naturforschende Gesellschaft, Zürich, Vierteljahresschrift* 56 (1911): 1–14. This work, among much else, was the origin of the famous Twin Paradox.

Einstein, Albert. "The Fundamentals of Theoretical Physics." In *Ideas and Opinions*. New York: Bonanza, 1954, pp. 323–335. This is the origin of Einstein's comment about Lorentz Invariance.

Einstein, Albert. "The General Theory of Relativity." In *The Meaning of Relativity*. Princeton, NJ: Princeton University Press, 1955. As a motivation for why time runs slowly in accelerated frames, Einstein describes an observer on a rotating disk. This work was originally published in 1921, 5 years after his major paper on General Relativity. I have expanded his analogy somewhat in my "derivation" of general relativistic time dilation.

Einstein, Albert. *Relativity: The Special and the General Theory*. Ed. Robert W. Lawson. New York: Three Rivers Press, 1961.

Einstein, A. "Über das Relativitätsprinzip und die aus demselben gezogenen Folgerungen." *Jahrbuch der Radioaktivität und Elektronik* 4 (1907): 411–462. Available in English: "On the Relativity Principle and the Conclusions Drawn from It." In *The Collected Papers*. Vol. 2. Trans. Anna Beck. Eds. John Stachel and Varadaraja V. Raman. Princeton: Princeton University Press, 1989, pp. 433–484. Though he wouldn't finalize general relativity for nearly a decade afterward, Einstein found certain key results, like gravitational time dilation, very shortly after discovering special relativity.

Einstein, A., H. A. Lorentz, H. Weyl, and H. Minkowski. "The Foundation of the General Theory of Relativity." *The Principle of Relativity: A Collection of Original Memoirs on the Special and General Theory of Relativity*. New York: Dover, 1952, pp. 109–164. This is a reprint and translation of Einstein's foundational work: "Die Grundlage der allgemeinen Relativitätstheorie" (1916).

Gott, J. Richard III, and Deborah Freedman. "A Black Hole Life Preserver." 2003. http://arxiv.org/abs/astro-ph/0308325. Gott and Freedman show that the period of time between being mildly uncomfortable and being ripped apart by a black hole is about 0.2 second. This is more or less independent of the mass of the black hole and holds up to masses of 10,000 times that of the sun. I've found it a nice problem to assign to grad students learning about General Relativity.

Haugen, Mark P., and Claus Lämmerzahl. *Principles of Equivalence: Their Role in Gravitation Physics and Experiments That Test Them*. New York: Springer, 2001.

Hawking, S. W. "Black Hole Explosions?" *Nature* 248, no. 5443 (1974): 30. The first paper on Hawking radiation.

Hawking, S. W. "Information Loss in Black Holes." *Physical Review D* 72, no. 8 (2005): 084013.

Isaacson, Walter. *Einstein: His Life and Universe*. New York: Simon & Schuster, 2007, p. 191. This is the origin of the quote (at least to me) regarding Einstein suggesting that astronomers should look for the gravitational lensing signal from GR during a total solar eclipse.

Norton, John D. "General Covariance and the Foundations of General Relativity: Eight Decades of Dispute." *Reports on Progress in Physics* 56 (1993):

791–858. Norton presents a very comprehensive discussion of the history of ideas in the grounding of general relativity, beginning with the Equivalence Principle.

Unruh, W. G. "Notes on Black-Hole Evaporation." *Physical Review D* 14, no. 4 (1976): 870. One of several derivations of several papers discussing and deriving the effect now known as Unruh radiation.

Wheeler, John Archibald, and Kenneth William Ford. *Geons, Black Holes, and Quantum Foam: A Life in Physics*. New York: Norton, 1998, p. 235. Wheeler made his comment about spacetime and curvature many times. However, this remains one of the pithiest versions of it.

CHAPTER 7

Bell, John. "On the Einstein Podolsky Rosen Paradox." *Physics* 1, no. 3 (1964): 195–200. Bell's Inequality (first described in this paper) was a method that could distinguish between the standard (Copenhagen) interpretation of quantum mechanics and Einstein's hidden variables. When the method was finally put to the test in the 1980s, the standard version of quantum mechanics won out. There were no hidden variables, and randomness and spooky action at a distance really seemed to be real.

Bennett, C. H., G. Brassard, C. Crepeau, et al. "Teleporting an Unknown Quantum State via Dual Classical and Einstein-Podolsky-Rosen Channels." *Physical Review Lett*ers 70 (1993): 1895–1899. This is the original theory paper describing the nature of using quantum entanglement for quantum teleportation.

Bernstein, Jeremy. *Quantum Profiles*. Princeton: Princeton University Press, 1991, p. 84. Origin of the "Einstein should have been right" quote.

Boschi, D., S. Branca, F. De Martini, et al. "Experimental Realization of Teleporting an Unknown Pure Quantum State via Dual Classical and Einstein-Podolsky-Rosen Channels." *Physical Review Letters* 80, no. 6 (1998): 1121–1125. This is the first experimental success in using quantum teleportation. In this case, a single photon was transported.

Dirac, P. A. M. "The Development of Quantum Mechanics." In *Conferenza Tenuta il 14 Aprile 1972*. Rome: Accademia Nazionale dei Lincei, 1974. Origin of Dirac's quote regarding indeterminacy.

Einstein, A., B. Podolsky, and N. Rosen. "Can Quantum-Mechanical Description of Physical Reality Be Considered Complete?" *Physical Review* 47, no. 10 (1935): 777–780.

Gilder, Louisa. *The Age of Entanglement When Quantum Physics Was Reborn*. New York: Knopf, 2008. A very nice discussion of the EPR paradox, Bell's inequality, and the tests of quantum mechanics.

Heisenberg, Werner. "Critique of the Physical Concepts of the Corpuscular Theory." In *The Physical Principles of the Quantum Theory*. Trans. Carl Eckhart and Frank C. Hoyt. Chicago: University of Chicago Press, 1930, p. 20. This is the origin of the "Every experiment destroys some of the knowledge" quote.

Heisenberg, Werner. "Über den Bau der Atomkerne. I." *Z. Phys.* 77, no. 1 (1932). Available in English: D. M. Brink. *Nuclear Forces*. Elmsford, NY: Pergamon, 1965.

Jin, Xian-Min, Ji-Gang Ren, Bin Yang, et al. "Experimental Free-space Quantum Teleportation." *Nature Photonics* 4, no. 6 (2010): 376–381.

Krauss, Lawrence Maxwell. *The Physics of* Star Trek. New York: Basic Books, 1995. An excellent discussion of many issues in modern physics, but most relevant for this discussion is quantum teleportation.

Krulwich, Robert. "Commemorate Caesar: Take a Deep Breath!" *Morning Edition,* National Public Radio, March 15, 2006. www.npr.org/templates/story/story.php?storyId=5280420.

Parfit, Derek. "What We Believe Ourselves to Be." In *Reasons and Persons*. Oxford, UK: Clarendon Press, 1984, chap. 10.

Wootters, W. K., and W. H. Zurek. "A Single Quantum Cannot Be Cloned." *Nature* 299, no. 5886 (1982): 802–803. The origin of the No Clone Theorem.

Yin, Juan, He Lu, Ji-Gang Ren, Yuan Cao, et al. "Teleporting Independent Qubits through a 97 km Free-Space Channel." http://arxiv.org/abs/1205.2024 (2012).

CHAPTER 8

Dirac, P. A. M. "A Theory of Electrons and Protons." *Proceedings of the Royal Society of London* A126 (1930): 360–365. In his derivation of antimatter, Dirac mistakenly (but quite reasonably) thought that the proton was the electron's antiparticle.

Feynman, Richard P. "Identical Particles." *The Feynman Lectures on Physics*. Vol. 3. Eds. Robert B. Leighton and Matthew L. Sands. Reading, MA: Addison-Wesley, 1963, p. 4–3. Feynman's (often quoted) comment about the difficulty in explaining the Spin Statistics Theorem.

Heisenberg, Werner. *Physics and Philosophy: The Revolution in Modern Science*. St. Leonard's: New South Wales, 1959. Lectures delivered at University of St. Andrews, Scotland, winter 1955–1956. Heisenberg notes: "We have to remember that what we observe is not nature herself."

Pauli, Wolfgang. "The Connection between Spin and Statistics." *Physical Review* 58 (1940): 716–722.

Pauli, Wolfgang. "Über den Einfluss der Geschwindigkeitsabhängigkeit der Electronenmasse auf den Zeemaneffekt." Available in English: "Zeeman-Effect and the Dependence of Electron-Mass on the Velocity." *Zeitschrift fur Physik* 31 (1925): 373. Pauli was awarded the 1945 Nobel Prize in physics for this work in which he first developed what is now known as the Pauli Exclusion Principle.

CHAPTER 9

Englert, F., and R. Brout. "Broken Symmetry and the Mass of Gauge Vector Mesons." *Physical Review Letters* 13, no. 9 (1964): 321–323.

Feynman, Richard P. *QED: The Strange Theory of Light and Matter*. Princeton, NJ: Princeton University Press, 1985.

Griffiths, David. *Introduction to Elementary Particles*. Weinheim, Germany: Wiley-VCH, 2008. An excellent advanced undergraduate-level textbook on the fundamentals of particle physics.

Griggs, Jessica. "Peter Higgs: Boson Discovery Like Being Hit by a Wave." *New Scientist,* July 10, 2012.

Guralnik, G. S., C. R. Hagen, and T. W. B. Kibble. "Global Conservation Laws and Massless Particles." *Physical Review Letters* 13, no. 20 (1964): 585–587.

Higgs, Peter W. "Broken Symmetries and the Masses of Gauge Bosons." *Physical Review Letters* 13, no. 16 (1964): 508–509.

"The Hunt for the Higgs Boson." *Science Scotland*. www.sciencescotland.org/feature.php?id=14. Interview with Peter Higgs that refers to his comment, "This summer I have discovered something useless."

"Latest Results from ATLAS Higgs Search." ATLAS Experiment, July 4, 2012. www.atlas.ch/news/2012/latest-results-from-higgs-search.html.

Lederman, Leon M., and Dick Teresi. *The God Particle: If the Universe Is the Answer, What Is the Question?* Boston: Houghton Mifflin, 1993. A great book with a truly dreadful title.

"Observation of a New Particle with a Mass of 125 GeV | CMS Experiment." CMS Public | CMS Experiment, July 4, 2012.

Oerter, Robert. *The Theory of Almost Everything: The Standard Model, the Unsung Triumph of Modern Physics*. New York: Pi, 2006. An excellent overview of the Standard Model in layman's terms.

Overbye, Dennis. "Physicists Find Elusive Particle Seen as Key to Universe." *The New York Times*, July 4, 2012. www.nytimes.com/2012/07/05/science/cern-physicists-may-have-discovered-higgs-boson-particle.html. One of many articles describing the Higgs field as a cosmic molasses.

Peskin, Michael Edward, and Daniel V. Schroeder. *An Introduction to Quantum Field Theory*. Reading, MA: Addison-Wesley, 1995.

Thomson, J. J. *James Clerk Maxwell: A Commemoration Volume, 1831–1931*. Cambridge: Cambridge University Press, 1931. Einstein's contribution described Maxwell's introduction of fields as being central to physics.

Weyl, H. "Eine neue Erwiterung der Relativitatstheorie." *Annalen der Physik* 59 (1919): 101. Weyl, in addition to his other contributions to symmetry, was also the first to realize that the assumption of local Gauge Symmetries could be used to derive electromagnetism.

Yang, C. N., and Mills, R. "Conservation of Isotopic Spin and Isotopic Gauge Invariance." *Physical Review* 96, no. 1 (1954): 191–195.

CHAPTER 10

Bacon, Francis. *Philosophical Studies: c. 1611–c. 1619*. Eds. Michael Edwards and Graham Rees. Oxford, UK: Clarendon, 1996. Bacon's description of Thales's philosophy is taken from his *De Principiis Atque Originibus*.

Diamond, Jared M. *Guns, Germs, and Steel: The Fates of Human Societies*. New York: Norton, 1998.

Dimopoulos, S., and H. Georgi. "Softly Broken Supersymmetry and SU(5)." *Journal of Nuclear Physics B* 193 (1981): 150–162. This paper is an adaptation of supersymmetry to a popular GUT model.

Feynman, Richard P. *QED: The Strange Theory of Light and Matter*. New York: Penguin, 1990, p. 128. This is the origin of Feynman's quote disparaging the renormalization technique.

Georgi, H., and S. L. Glashow. "Unity of All Elementary Particle Forces." *Physical Review Letters* 32 (1974): 438–441. This is the original paper proposing the SU(5) grand unification scheme.

Goenner, Hubert F. M. "On the History of Unified Field Theories." *Living Reviews in Relativity* 7 (2004): 2. www.livingreviews.org/lrr-2004-2.

Lisi, A. G. "An Exceptionally Simple Theory of Everything." 2007. http://arxiv.org/abs/0711.0770.

Lisi, Garrett. "Garrett Lisi: A Theory of Everything." TED Conference presented February 2008. www.ted.com/talks/garrett_lisi_on_his_theory_of_everything.html.

Lisi, G., and J. O. Weatherall. "A Geometric Theory of Everything." *Scientific American* 303, no. 6 (2010): 54–61.

Newton, Isaac. *The Principia.* Trans. I. B. Cohen and A. Whitman. Berkeley: University of California Press, 1999. Translation of Newton's Third Law of Motion.

Peacock, John A. "The Standard Model and Beyond." In *Cosmological Physics.* Cambridge: Cambridge University Press, 1999, pp. 216–270.

Susskind, Leonard. "New Revolutions in Particle Physics: Supersymmetry, Grand Unification, and String Theory." Lecture for Stanford Continuing Studies Program. 2009. http://itunes.apple.com/us/itunes-u/supersymmetry-grand-unification/id384233338.

"WMAP Recommended Parameters Constraints." LAMBDA—Legacy Archive for Microwave Background Data, June 21, 2011. http://lambda.gsfc.nasa.gov/product/map/current. The values for cosmological parameters are a constantly moving target, which is why you'll read slightly different numbers in different sources. They are measured with error bars, of course, but the Wilkinson Microwave Anisotropy Probe (WMAP) team has some of the best combined estimates of the parameters.

ACKNOWLEDGMENTS

Writing a book is fun, but you know what? It's also hard work, and I wouldn't have been able to get through it without the love, support, and feedback of a lot of people. First and foremost, my wife, Emily Joy, has read every line and has given me the tough love that I so desperately need. I thank my family and (though they didn't give me any comments at all) my daughters, Willa and Lily.

Thanks to all of my friends and colleagues who provided comments and feedback on various parts of the draft and useful conversations: Sean Carroll, Rich Gott, Richard Henretta, Andy Hicks, Lynn Hoffman, Josh Kamensky, Adrienne Leonard, Sean Lynch, Kate Mason, Kevin Owens, John Peacock, Dawn Peterson, Tina Peters, Som Tyagi, Liz Fekete Trubey, and Enrico Vesperini.

I also want to acknowledge Annalee Newitz, Charlie Jane Anders, and everyone at io9.com. Annalee and Charlie Jane were very supportive throughout, and many of these topics were road tested at io9. I would also like to thank all of the readers whose comments helped identify roadblocks to comprehension (theirs and mine).

Not every academic environment smiles on popular science writing as a serious pursuit. I'd like to thank my department head, Michel Vallieres, and my dean, Donna Murasko, for taking the larger view of scholarly work.

I'd like to thank my agent, Andrew Stuart. Andrew did more than simply sell this book; he was an excellent sounding board and advocate and was instrumental in helping me find my voice. My editor, Stephen Morrow, and his outstanding assistant, Stephanie Hitchcock, were tough but fair with their virtual red pen, and in stemming my natural verbosity (and trimming some of the more terrible jokes), they improved the book immensely. Thanks also to my amazing illustrator, Herb Thornby, who so deftly and beautifully conveyed ideas that my words may have otherwise obfuscated.

I also gratefully acknowledge Michael Blanton and the Sloan Digital Survey Collaboration for use of their data.

INDEX

Note: Page numbers in *italics* indicate photographs and illustrations.